High rupturing capacity fuses

High rupturing capacity fuses

DESIGN AND APPLICATION FOR SAFETY IN ELECTRICAL SYSTEMS

E. JACKS *C.Eng. F.I.E.E.*

Commercial Director, GEC-English Electric Fusegear Ltd

LONDON

E. & F.N. SPON LTD

First published 1975
by E. & F. N. Spon Ltd
11 New Fetter Lane, London EC4P 4EE

© *1975 E. Jacks*

Typeset by Preface Ltd., Salisbury, Wilts,
and printed in Great Britain by
Lowe & Brydone (Printers) Ltd
Thetford, Norfolk

ISBN 0 419 10900 5

Distributed in the U.S.A.
by Halsted Press, a Division
of John Wiley & Sons, Inc., New York

Library of Congress Catalog Card Number — 74-8613

Preface

The chapters in this book have previously been published as articles or conference and lecture papers. Details of the date of publication or of the conference concerned are given in a footnote at the beginning of each chapter.

Thanks are due to the following for permission to reprint the chapters indicated: The Manchester Association of Engineers for chapter 1, GEC–English Electric Fusegear Ltd for chapters, 2, 4, 6, 9, 10, 12, and 14, the Institution of Electrical Engineers for chapters 3, 7, 11 and 15, the I.E.E.E (U.S.A) for chapter 13, *Electrical Review* for chapters 5 and 16, and *Electrical Times* for chapter 8.

E.J.

Acknowledgements

The author acknowledges his debt to GEC–English Electric Fusegear Ltd., and to The English Electric Co. Ltd., which preceded it, for permission to use material and data relating to their products.

These Companies have led the world in fuse development, manufacture and application for almost 50 years and enjoy a reputation in this field which is recognized and respected by electrical engineers everywhere.

E.J.

Contents

Introduction

The papers and articles collected in this volume were published separately over a number of years. Each one was written to deal with contemporary problems or as a review of contemporary practice at the time of the original publication, and stands as an exposition of a particular aspect of fuse technology and practice.

Many of the problems discussed continue to recur and are likely to do so in the future, which may explain the steady and sustained demands for reprinting at regular intervals. Collectively, the papers now provide a record of the sequence of events in the field of fuse protection over a period of twenty years or more. During this time electrical distribution systems have undergone a number of substantial changes through which new techniques and practices have evolved. A knowledge of these processes is necessary for the understanding of present day systems and of the standards, ratings and regulations which govern them. Protective devices and fuses in particular need to be viewed in the light of these changes because present standards and methods must necessarily take account of those which preceded them. At any given time there may be several generations of development still in service. It is seldom possible to make a completely new departure from existing conditions.

Changes in the field of fuse protection have followed in quick succession because of the rapid growth in energy utilization but perhaps more importantly because the versatility of fuses has enabled them to fulfil an ever widening range of duties. These stem from the introduction of new materials and techniques in system construction and the economic pressures which have accentuated the process. Almost every move towards efficiency and economy makes the task of protecting the system more critical. This has placed greater demands on fuse protection and has in consequence stimulated advances in fuse design. H.R.C. fuses have now reached a highly sophisticated form, and are used in situations which would not have been considered possible even a decade ago.

Fuses and in particular H.R.C. fuses are in vast and wide-spread use throughout the world, occupying an almost indispensable role in the protection of electrical systems. Their function in providing for the safety of plant and the people who operate it gives them a very considerable significance in the economics of electrical distribution.

Published information in this field has been rather scarce, probably for the reason that there are relatively few specialists in fuse technology compared with the great number of engineers who are concerned with fuses in practice. The appetite for information has increased in direct proportion both to growth in terms of usage and to the advances in fuse technology.

Although the H.R.C. fuse is relatively simple in construction it nevertheless has a very complex function. It must respond to the transient behaviour of the system which it protects under fault conditions. Since there is a huge variety of systems and since transient phenomena are equally complex and varied there is an almost infinite range of conditions to be satisified.

The interruption of a high power fault involves the containment by the fuse of large values of fault energy. This manifests itself in arcing within the fuse and must be efficiently and safely controlled. The fuse must perform with complete safety whenever it is called upon to do so even after a life of 30 years or more in service. Its performance in given circumstances must be accurately known and completely predictable.

These papers and articles are mainly concerned to show that fuse designs have kept pace with contemporary requirements and to give practical proof of this contention. Practising engineers need this information to substantiate and qualify the more formal specifications and performance data which they use in day to day operations. The fact that such data are often taken for granted by unqualified operators is perhaps a tribute to their validity, but qualified engineers need to know the bases of rating and performance and to appreciate the scientific principles involved.

The responsibility for operating and maintaining electrical services increases as industry becomes more dependent upon them. This is particularly so as industry becomes more 'capital intensive'. Safety and reliability assume greater importance in these cases because the risks in human and financial terms increase. Perhaps the main value of this book is that it may contribute to a better understanding of the problems involved in the assessment and calculation of the risks and so encourage the highest standards of safety in given situations.

Ormskirk   E.J.
May 1974

1 Electrical power distribution in factories

The Manchester Association of Engineers, to which this paper was presented in 1951, is one of the oldest engineering societies of its kind in the world, having been founded early in the 19th century. Its earlier members included many distinguished engineers, who from the very beginning recognized their responsibility to engineering and the community at large. Immediately after the Second World War, the Association concerned itself with the reorientation of engineers from war service to peace-time industrial practice. It was in this context that the paper was requested and written.

The illustrations and commentary provide an interesting record of contemporary practice over a range of industries but the principles set out are valid for any era.

Emphasis in electrical protection at this time was on rupturing and breaking capacity. System capacity had grown enormously during the war years without any corresponding opportunity to uprate switch or fusegear to meet the higher fault levels. Fuses and fusegear were rapidly developed to meet the need. New testing facilities and techniques were brought into being and strenuous efforts were made to educate operating engineers to appreciate the problems and the means of overcoming them.

Originally published in October 1951 by The English Electric Co. Ltd., Fusegear Division, with the permission of The Manchester Association of Engineers.

1.1 Introduction

The subject of electrical distribution in factories is obviously too wide to be dealt with in one paper. It has been necessary to restrict this paper to those aspects which the author feels will have the widest appeal to readers.

Generally speaking, the paper is confined to the consideration of distribution equipment, with particular reference to the application of modern H.R.C. (high rupturing capacity) fuses, but it is emphasized that these must be considered in relation to all the other equipment, such as cables, fittings, motors and other consuming devices, which comprise the modern factory installation.

The paper, therefore, consists of an objective examination of the basic principles of electrical distribution followed by some discussion concerning the choice of equipment and concluding with a consideration of problems affecting a number of specific industries.

1.2 The fundamental principles of industrial electrical distribution

The first problem in conducting an examination of the fundamental principles of electrical distribution is to define the principles themselves. In practice it is a fact that no two factories are alike and there is always ample scope for individual preference, personal taste, and technical ingenuity among the many alternative schemes which present themselves. In spite of this, all good installations should conform to certain fundamental requirements of adequacy, safety, reliability, adaptability and, last but not least, economy. The manner in which these requirements are set down is quite arbitrary but for the purpose of this paper it is convenient to consider them under the following headings.

1. The installation should provide an adequate supply of electrical energy at any point where it is required or where it may be required in the immediate future.

This statement is a fairly obvious one but it contains the essence of all that matters most to the people whom the installation serves.

The distribution of electrical energy is a means to an end, namely, to provide the power to produce goods or perform services. It is the purpose of the distribution system to carry the supply of energy to the places where it is required at the time it is required. If it cannot do this fully, either due to lack of capacity or avoidable breakdown, it is wasteful both of capital and of production, because the plant it serves cannot be used to maximum efficiency. The losses due to such causes can sometimes far exceed the capital cost of the electrical distribution scheme itself and there is no doubt that it pays to consider very carefully the method of distribution to be adopted as well as the design of the equipment it comprises.

It is advisable in all but the smallest jobs to prepare detailed drawings and to plan the installation on paper before beginning the actual work of installing the equipment. Only in this way can the installation be planned to do its job efficiently while ensuring the minimum voltage drop and thermal loss within the accepted limits of tolerance, with the minimum overall amount of copper. The first step in planning should always be to obtain detailed information of the magnitude and layout of the load, that is, the job should always start by considering each circuit from the point where the motors, etc., are installed. The layout can then be properly sectionalized by grouping a number of individual circuits and feeding from a section switchboard or fuseboard. Each section can then be fed back from the main switchboard by a single cable. The size of the sections in terms of current rating can be calculated fairly accurately and from this information the size and shape of the main switchboard can then be decided.

The selection of the most suitable distribution scheme depends on many factors, including the nature and source of the supply, the size of the load, the load factor, the geographical layout of the site, the relative permanence of plant, and the possibilities of extension. The following typical examples will serve to illustrate briefly some of the various alternatives available but it will be obvious that these are only a few of the many combinations possible.

Small factories taking loads up to 150/200 kW are normally fed by public supply direct from the network. A small switchboard is used to distribute the incoming supply to the various circuits which branch out radially from the busbars.

For large factories the supply may be taken from a transformer substation fed from the H.T. supply by a single H.T. feeder. For the L.T. distribution a radial scheme may again be employed or, alternatively, for certain load conditions a ring system may be an advantage. Where more than one transformer is required suitable bus coupling between the L.T. busbars ensures maximum availability of supply. In cases where the manufacturing processes of the factory are continuous and cannot be stopped for more than a very short time, duplicate H.T. feeders are usually provided.

Large factory sites or large factories with heavy loads require several substations which may be fed from an H.T. ring or radial scheme. The H.T. supply may be fed from single or duplicate feeders again according to the degree of availability required. In some cases it is beneficial to interconnect on the L.T. side as well.

Having decided the source of supply and the general shape of the main switchboard, the actual method of feeding the various load concentrations can be chosen. The most common method is the radial system which employs main distributor cables branching out individually from the main switchboard to section distribution boards and then through smaller distribution boards to the motor starters or other switches directly controlling the load. Occasionally a ring main is useful but only for certain types of load. Ring mains should not be used unless the advantage from them is quite clear. Although it is possible to save copper by this means in certain instances, it is more likely that the loss of discrimination will outweigh the saving. There is evidence that ring cabling is not now so fashionable as it used to be, except in the domestic field where the type of load is eminently suitable for it.

Whatever type of system is chosen, provision for future extension should always be seriously considered. Ample capacity should be allowed at the main switchboards and in the main distribution cables. Spare fuse-ways should be left in every distribution board and steps should be taken to keep a continual check on the system loading.

Most installations nowadays employ the standard voltage of 415/240 at 50 cycles on a three-phase, four-wire system. While most power loads are three-phase, three-wire balanced loads, the need for single-phase to neutral supplies often occurs. It is to be advocated that the neutral of the system is taken to each of the main section boards so that 240 V equipment can be connected without undue difficulty. This question has to be weighed very carefully at the present time, in view of the high cost of copper, but under normal conditions there is no doubt that the neutral should be run to the main section boards.

Lighting should be segregated from the power loads in order, when necessary, to enable maintenance to be carried out on the power circuits during the hours of darkness. Also, where an installation includes one or two comparatively large motors among a greater number of small ones, it is desirable to run a cable for each large motor right back to a main distribution point. This gives greater control and facilitates protection especially from the point of view of discrimination.

2. **The installation should be capable of withstanding and interrupting any electrical fault which may occur and should discriminate in such a manner that the interruption in supply is localized to the faulty circuit only, leaving the healthy circuits unaffected.**

The importance of this requirement cannot be overemphasized. Almost any kind of installation will carry electrical energy while stable conditions exist but the test of a good and well-designed installation is in its ability to withstand fault conditions without suffering damage, except that caused at the actual point of the fault.

As in most other cases, whether technical or personal, prevention is better than cure, and the first measure of protection is in the quality of the equipment comprising the installation. Breakdowns are costly things, especially in these days of accelerating production and the first consideration must always be to prevent faults occurring as far as is humanly possible. Unfortunately, faults do occur even on the best installations and protective measures are necessary; but these should be called upon to deal only with a minimum of unavoidable faults.

It is convenient to regard faults under two general headings:

(a) overcurrents due to deliberate or accidental overloading;
(b) short-circuit faults.

The former may be regarded as referring to fault currents of a magnitude up to several times the rating of the connected apparatus, while the latter can refer to currents above the overload values up to the maximum short-circuit level of the system. Earth faults, as distinct from earth leakage, can normally be dealt with as short-circuit faults or as overload faults, depending upon the point on the system at which the fault occurs. In the case of overloads, the offending circuits are usually disconnected by motor starters or similar devices actuated by either a thermal or magnetic device. Such devices or contactors designed for industrial use are almost invariably of low breaking capacity, that is to say, they are designed to interrupt currents of the order of ten times their nominal continuous current rating and must be so arranged in the circuit that they are never called upon to interrupt currents in excess of this. This can easily be achieved by backing them up with H.R.C. fuses either fitted integrally with the contactor or elsewhere in the circuit.

For short-circuit protection the modern H.R.C. fuse is now by far the most widely used device both in this country and overseas. Circuit-breakers are also used in certain instances, but their cost precludes their use except when employed for special reasons, Fig. 1.1 shows a simplified schematic diagram of a typical installation in which the supply is taken from a transformer through some form of main switch to the consumer's busbar. The main circuits are taken from the busbar through section switches to section distribution boards and then through the final distribution boards to the ultimate loads. Fuses, or other protective devices, are inserted at each of these stages, the simple rule being that protection is necessary wherever the conductors reduce in cross section.

Except in certain special cases, the only position in a factory installation where overcurrent protection is necessary or desirable is in the control gear adjacent to a motor or other similar apparatus. The cables and distribution equipment may be regarded as being liable only to short-circuit faults, and it is usual nowadays to apply short-circuit protection only. The provision of ammeters in appropriate positions where they can be easily seen by the maintenance or operating staffs is usually sufficient to prevent overloads.

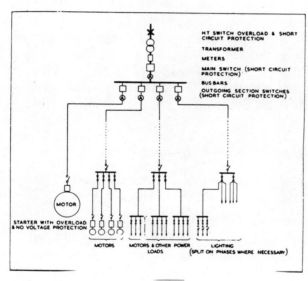

FIG. 1.1
Schematic diagram of theoretical circuit

Examining Fig. 1.1 again, it will be obvious that any protective device attached to the main switch will clear faults in the busbar zone only. In practice it is found that this type of fault is extremely rare in good-class equipment and, in any case, can be cleared by the H.T. circuit-breaker. The cost of automatic protection at this point cannot always be justified in view of these considerations. It is obligatory, of course, to provide some means of disconnecting the supply to conform to the Safety Regulations of the Factories Act. This may take the form of a metal-clad, gang-operated isolator in the simplest case, or a fully 'on-load' fuse switch, or in certain circumstances an automatic circuit-breaker. In cases where doubt exists, the advice of the Factory Inspector may be sought, but it is often found that for both technical and economic reasons the simplest form of main switch will suffice.

Automatic circuit-breakers (which, incidentally, must be of adequate breaking capacity) are used where remote operation or automatic on-load changeover is necessary and, in some cases, as a precaution where trained operating staff are not available. It is perhaps also true to state that they are sometimes used for no reason at all except to preserve an old custom.

The section switches outgoing from the main busbars are also required to provide short-circuit protection, and fuse switches with 'on-load' features are again the logical choice.

FIG. 1.2
Typical main switchboard with 1500 Amp incoming isolator

The section fuse boards, and sometimes the final distribution boards, may or may not be provided with isolators. In the author's view where the distance between the main board and the distribution board is relatively great, or where the two are in separate rooms, an isolator is desirable to facilitate maintenance and to ensure maximum safety. It is recognized, however, that many engineers achieve these objects by alternative means. If a switch is fitted to control a fuse board in this manner, fuses should not normally be fitted in the switch itself as this would only duplicate the fuses at the incoming end of the distributor and would complicate matters in the event of a fault occurring across the distribution board busbars.

The modern H.R.C. cartridge fuse is now so prominent in the protective scheme of all modern industrial installations that an appreciation of its characteristics is now essential to the factory engineer. It is, therefore, proposed to deal with this particular subject under a separate heading later in the chapter.

3. **The installation should be both electrically and mechanically sound and fully accessible for easy maintenance so that deterioration cannot take place to the detriment of production.**

The best way of ensuring that an installation is electrically and mechanically sound is, of course, to use equipment which has been specifically designed for industrial use and which has acquired a reputation in this field. Although the examples

FIG. 1.3
Distribution board with isolator

discussed within the terms of this chapter apply only to the fusegear equipment, this statement is equally true for all other components and in this respect it is well to remember the old adage that 'the strength of the chain is in the weakest link'.

From an electrical point of view it is necessary to ensure that all the conducting parts are able to carry the current required without overheating and to be certain that the insulation is good.

Switches which are required to 'break' load should be capable of doing so without undue deterioration of the contacts and the possibility of switches having to 'make' and 'break' on overcurrent due to a stalled motor, etc., should not be overlooked.

TABLE 1.1
Example of breaking-capacity tests

(Actual results of tests on 300 A T.P. switch fuse)
System: 415 V, three-phase, 50 cycles
Power factor of test: 0.8 lagging
Time interval between tests: not exceeding 0.5 sec

Test 1. To test severest
service conditions (fitted with
300 A H.R.C. cartridge fuses)

Test 2. To determine
ultimate breaking capacity
(fitted with solid copper links
in lieu of fuses)

Number of make/break operations	Current (A) (r.m.s.)	Number of make/break operations	Current (A) (r.m.s.)
10	460	10	1830
10	714	5	2130
10	942	5	2390
10	1180	5	2760
10	1430	5	3060
		5	3420
		5	3770
		5	3990
		4	4410

Note: Fuses operated when
current was increased to
2160 A; contact surfaces
undamaged.

Note: No flashovers occurred;
switch contacts welded-in on
fifth operation at 4410 A.

Connections should be as good as it is possible to make them. Contact surfaces should be large enough to prevent a rise in temperature above that of the conductors surrounding them, and clamping bolts, lugs and pinching screws, should be large enough to stand tightening by substantial tools. Wherever possible, connections should be soldered, and clamped connections should be secured by high-tensile steel bolts which should be chosen to avoid the necessity of using special tools to fit them. Connections to busbars are usually best made in the form of clamps. Drilling of busbars is not the best of practice as it reduces the cross section of the busbar without necessarily giving the best connection and it considerably reduces the adaptability of the gear. Incidentally, connections should always be checked on site, both before the gear it put into commission and periodically thereafter.

A cardinal rule in design is that the number of connections of any kind for a given circuit should be kept to an absolute minimum consistent with practicability.

Although the current-carrying capacity of a piece of equipment depends largely on its general design, it must be remembered that its performance will also be governed by the circuit conditions and other local factors. The choice of a switch, for instance, will be influenced by the type of duty it is expected to perform, and

FIG. 1.4
Example of effective insulation and shrouding of live parts

such factors as ambient temperature and humidity, pollution (both atmospheric and
due to spillage), frequency of switching, load and diversity factors, and opportunity
for maintenance including availability or otherwise of maintenance staff, will require
due consideration. The relative importance of the circuit controlled by the switch
should also receive due consideration as it is always prudent to be fairly generous
with the current-carrying capacity of equipment controlling continuous processes, etc.

Insulation should have mechanical strength appropriate to the duty of the
equipment as well as having the necessary dielectric properties. The shrouding of all
live metal is to be advocated wherever possible. Creepage distances should be
adequate; non-tracking materials should be employed and in cases where air
insulation is unavoidable, air space between parts of different potentials should be
generous enough to cope with the worst atmospheric conditions likely to be
encountered.

Mechanical factors involved in the design of industrial equipment are closely
related to the electrical factors and must satisfy the same general requirements. By
way of example, it is useful to consider the difference between industrial equipment
and, say, domestic equipment. Generally speaking, the former is in more constant
use, it receives rougher handling, the consequences of failure are economically more
serious and it is subject to more rigorous official safety regulations. The mechanical
design of equipment for factory use should, therefore, be appropriate to the duty
and is worthy of some thought. Static equipment should be robust enough to
withstand the effects of vibration and the rigours of factory traffic. Moving parts
should be able to withstand normal wear and tear without strain or failure. The
designer of such equipment must give due attention to the correct choice of
materials, as, for example, hard steels to avoid wear, brass bearing surfaces to avoid
seizures and compatible metals to avoid corrosion. All these points, of course, come
within the scope of manufacturing technique and experience, both of which are
essential to a high degree in the production of good-class equipment.

Lastly, the mechanical design should be such that all equipment is fully accessible for maintenance and cleaning. Small but important points, such as captive studs and nuts, and assemblies which can be taken apart for cleaning and refitted without difficulty, will all add up to the effectiveness of the installation and may often make the difference between a good job and a maintenance electrician's nightmare.

4. The installation should be pleasing to the eye and at the same time should be simple in layout so that the controlling switches and protective devices can be easily identified with the machinery they control.

The general trend towards improving factory conditions to attract the right type of labour has led in recent years to a more active consideration of the shape and appearance of electrical equipment in factories. Appearance is, of course, largely a matter of taste, but there is little doubt that equipment which is fundamentally sound from a functional point of view is also the best aesthetically. Where the technical and aesthetic requirements can be reconciled, this should, of course, be done. A case in point is the finish adopted on equipment. This can be both pleasing to the eye and really effective in preventing deterioration and corrosion.

The position of distribution equipment can sometimes be chosen to enhance, rather than detract from, the appearance of a workshop and some thought about the relative spacing of switchboards and distribution boards is always worth while.

Where possible, distribution equipment should be positioned adjacent to the load it controls in such a way that it can be easily identified with the connected circuits, thus ensuring the minimum of difficulty to personnel who may have to find a fault or disconnect a particular circuit for maintenance purposes. The proper labelling of equipment is another essential requirement in this connection. Labels should in every case be both substantial and legible.

The appearance of any installation is greatly enhanced if kept properly cleaned. The design of equipment, therefore, should be such as to facilitate this condition.

5. The installation should provide for the connection of additional loads without danger of overloading the system or creating unsightly appendages to the existing scheme.

As mentioned previously, it is always prudent to allow a good margin on the capacity of main busbars and main distributors and to allow spare fuse-ways wherever possible in distribution boards. Throughout industry electrical load is growing at a considerable rate and this trend is common to all factories almost without exception. Even the most seemingly static processes are found to require more and more kilowatts because of the increasing use of electricity for auxiliary services.

When laying out the main cables, etc., in a factory, a considerable proportion of the cost is in labour charges due to the cutting of trenches and holes through walls and the fixing of brackets and other supporting structures. These labour costs are to some extent independent of the size of cable being run and it follows that any extra margin allowed on the current carrying capacity of the scheme will not cause a proportional increase in the overall first cost.

If the installation is not adequaté to allow for normal expansion, the only two alternatives are to run completely new distributors at considerable cost or to succumb to the temptation of hooking temporary circuits from existing distributors. The latter course is dangerous to personnel, upsets the protection of the initial part of the installation, introduces considerable fire hazard, and lowers the general standard of the factory by its unsightly appearance. There is, in fact, nothing that can be said in favour of this practice which, unfortunately, is widespread enough to require emphatic mention in any paper which purports to advocate means for the betterment of engineering practice.

6. The installation should be completely safe and should conform in every way to accepted standards.

It is obligatory in this country to comply with the Safety Regulations of the Factories Act, and although these regulations are very general in nature, the co-operation between H.M. Inspectors and Industrial Engineers, has resulted in this country in standards of practice which are second to none in the world.

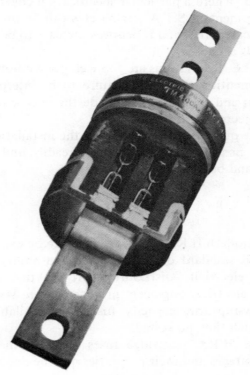

FIG. 1.5
*H.R.C. non-deteriorating cartridge fuse-link cut away
to show position of elements (filling removed)*

The interpretation of the Factories Act varies according to conditions and for general guidance the accepted authority on good wiring practice is to be found in *Regulations for the Wiring of Buildings* issued by the Institution of Electrical Engineers. Although these regulations are not primarily applicable to industry, they are usually accepted as a guide for this purpose. It is recognized, however, that they cannot possibly cover every contingency met with in practice as is evident from the fact that they are amended at frequent intervals to keep pace with modern requirements.

So far as the standards of equipment are concerned, the standard specifications issued by B.S.I. now cover a very wide range so that the user can safeguard himself by insisting upon the adherence to these standards by manufacturers of equipment.

An admirable example of the value of British Standard Specifications is to be found in B.S. 88: 1947, covering fuses. The provisions governing the conditions for short-circuit testing in this specification are particularly interesting in that they form an authoritative basis on which testing can be done by a body independent of the manufacturer. The Association of Short-Circuit Testing Authorities will only issue a Certificate of Rating for equipment which is tested strictly in accordance with the requirements of the appropriate B.S. An A.S.T.A. certificate is, therefore, the user's most reliable safeguard. Where a particular specification covers several categories for the same type of equipment, A.S.T.A. certificates can be procured for the lower as well as for the higher categories, and it behoves the user to be sure that the category appropriate to his needs is chosen.

The Factory and I.E.E. Regulations are too well known nowadays to need further comment, except to mention that they require to be interpreted in the 'spirit' in which they are written and should not be made the excuse for bad practice where their application is subject to controversy.

One last word on the question of safety is that the installation should be as simple and uncomplicated as possible so that it can be readily understood by the people who have to maintain and operate it.

1.3 H.R.C. fuses

The popularity of the modern H.R.C. cartridge fuse can be explained simply as being entirely due to its high standard of performance and ability to meet the requirements of present-day electricity distribution. At the time the H.R.C. fuse was introduced in Britain, electrical engineers in general were losing confidence in the types of rewirable low-rupturing-capacity fuses then available on account of the many disadvantages which they possessed.

The best designs of H.R.C. cartridge fuses have for many years completely overcome these disadvantages and their properties may be described as follows:

(a) *Rupturing Capacity.* The consequences of a short-circuit under modern network conditions can be very serious unless properly controlled, because of the enormous amount of power in the form of generator capacity which is now available on the mains. A 1500 kVA transformer with five per cent reactance can have a nominal fault level of some 30 MVA at its low-tension terminals. This is a conventional way of saying that some 40 000 r.m.s. amperes or approximately 100 000 assymetrical amperes could flow in a short-circuit near these terminals. In practice these values are somewhat reduced by the limit of the potential power on the H.T. side and by the impedance of the L.T. circuit but can, nevertheless, be sufficiently near to these figures to require protection up to the nominal values.

It must be borne in mind that it is impossible to know just how near to the transformer a particular fuse may be situated in practice, and it would obviously be impracticable to have fuses of different rupturing capacities for different parts of the circuit. Fig. 1.6 shows a typical oscillogram of a certified short-circuit test showing that standard fuses of this particular design can satisfactorily interrupt short-circuits of 35 MVA. For special cases where greater fault levels obtain, fuses of higher rupturing capacity are available.

(b) *High-Speed of Operation.* Unlike circuit-breakers, the H.R.C. fuse interrupts the short-circuit current long before its maximum value is attained. This cut-off effect greatly reduces both thermal and magnetic stresses on the equipment protected by the H.R.C. fuse. Fig. 1.6 shows this effect quite clearly and demonstrates that on severe short-circuits the fault is interrupted well within the first quarter of a cycle.

(c) *Non-Deterioration.* This means that all characteristics are maintained throughout the life of the fuse. The majority of fuses are expected to carry load for many years without attention. They are so unobtrusive, in fact, that they are sometimes forgotten and it is only when a fault occurs, causing them to 'blow', in some cases many years after they have been installed, that attention is focused on them. It is most essential that they should preserve their characteristics throughout their useful lives and so give efficient protection no matter when the fault occurs. This also implies that they should not 'blow' inadvertently when carrying normal load currents as so often happens with a rewirable fuse which is prone to fail due to oxidation and reduction of cross-sectional area. An essential factor in producing fuses to give these properties is the hermetic sealing of the silver element within the fuse body. In the case of the fuse illustrated, this is ensured by special cementing and the soldering of the end-caps. This policy has proved itself so successful that fuses which have been removed from circuit after approximately twenty years have been found to have had their original characteristics absolutely unimpaired.

(d) *Low-Temperature Operation.* This is necessary not only to eliminate the deterioration of the fuse itself but to prevent overheating of associated contacts. The fuses shown in Fig. 1.5 employ fabricated elements of pure silver which are specially designed to give a low temperature rise when carrying their full-rated current.

(e) *Accurate Calibration.* The fuses are accurately manufactured so that calibration curves can be published and issued to users. Operation within ±5 per cent of these curves can be obtained.

(f) *Accurate Discrimination.* The advantage of this feature is that on a distribution system protected by H.R.C. fuses, only faulty circuits are disconnected under fault conditions leaving healthy circuits unaffected. This is a most important feature in relation to production efficiency. It is possible because of the characteristics enumerated above and is the basis on which many distribution systems can be planned.

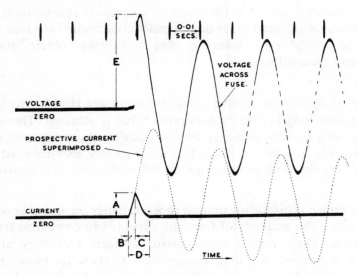

'ENGLISH ELECTRIC' CARTRIDGE FUSE LINK LIST No. TKF 300.

RMS SYMMETRICAL CURRENT	47,600 AMPS		CUT OFF CURRENT	29,200 AMPS	A
RMS ASYMMETRICAL CURRENT	72,500 AMPS		PRE ARCING TIME	0·00225 SECS	B
PEAK ASYMMETRICAL CURRENT	102,500 AMPS		ARCING TIME	0·00435 SECS	C
APPLIED VOLTAGE	440 VOLTS		TOTAL OPERATING TIME	0·0066 SECS	D
POWER FACTOR	0·13		ARC VOLTAGE MAX	602 VOLTS	E

FIG. 1.6
*Oscillogram of certified short-circuit test on
H.R.C. fuse*

The most important advantage which a well-designed cartridge fuse has over other types of fuse and over most other circuit interrupters is that the time of operation varies inversely as the prospective short-circuit current over a much wider range of fault conditions. That is to say, that, within practical limits, while the values of prospective short-circuit current increase, the time of operation will continue to decrease without reaching a 'definite minimum'. This means that a fuse of low current rating will 'blow' before a fuse of a higher rating no matter how heavy the fault. Recent tests have proved this to be true for a complete range of fuses up to currents of approximately 120 000 r.m.s. symmetrical amperes, a value which will satisfy almost any condition in industrial practice. It is necessary when planning the installation from the point of view of discrimination to use fuses of the same design and characteristic throughout. This will ensure that the time/current characteristic of each succeeding current size will not cross, but will tend to keep parallel up to the maximum values of fault current. Reputable manufacturers issue these curves freely to allow the user to satisfy himself on this point.

These properties of accurate discrimination are valuable and the extra care needed to lay out the installation network to take full advantage of them is time well spent.

(g) *Voltage Surge due to Interruption of Circuit kept well within Safe Limits.* This again is illustrated in Fig. 1.6. Whenever a short-circuit is interrupted, a voltage surge is liable to occur and may vary in magnitude according to the magnitude of the short-circuit and the circuit constants. A cartridge fuse can, however, be designed to control these overvoltages and keep them within safe limits. This is achieved by careful design of the element and by ensuring a high degree of consistency in manufacture. A potential cause of secondary breakdown can thus be avoided.

From one point of view the arc voltages, as these overvoltages are conventionally called, influence the design of the fuse body in that the creepage distance on the fuse itself must be sufficient to cope with the voltage surge resulting from the interruption of the most highly inductive circuit likely to be encountered. The voltage surge from the interruption of a resistive circuit is unlikely to be as severe but, again, it would obviously not be possible to produce different types of fuses for different circuit conditions.

(h) *Low Cost Compared with Similarly Rated Circuit Interrupters.* Where an alternative to the cartridge fuse is used as a short-circuit device, it must necessarily be of comparable rupturing capacity and circuit-breakers of 30 MVA or thereabouts are expensive items. It is not unusual, therefore, where circuit-breakers are necessary for other reasons, to employ a circuit-breaker of low rupturing capacity backed up by H.R.C. fuses.

Incidentally the advances which have been made in improving the characteristics of cartridge fuses have not meant any increase in the value of materials used, the improvements have thus been achieved without any increase in cost to the user. This applies, for instance, to the recent upgrading of one particular design from 25 to 35 MVA.

1.4 The choice of distribution equipment

Apart from the general considerations which have been mentioned, there are many factors affecting the choice of distribution equipment for particular applications.

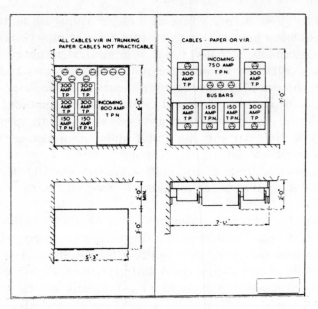

FIG. 1.7
*Diagram showing relative space taken by two forms of
switchboard*

Due to the ever-increasing use of electricity in both new and existing factories, the question of space factors under modern conditions is becoming more and more important. Space costs money or, in other words, the space taken up by electrical distribution equipment is space which cannot be otherwise used for production. Therefore, it is necessary when considering the choice of such equipment to purchase that which occupies the minimum space without sacrificing other essential features. Fig. 1.7 shows a comparison between two types of switchboard of similar ratings and illustrates clearly the amount of space which can be saved if due consideration is given to the problem. If the advantages of smaller gear can also be combined with improved methods of cabling and better operational features, then the proposition can be further enhanced.

Not all installations are in new factories, of course. In fact, it is probable that the majority of distribution equipment at present being manufactured is for installation in existing factories either to convert them to electric drive or to extend the existing electrification. Under these circumstances it is essential that the equipment should be fully adaptable and capable of easy extension. It often occurs that switchboards and other distribution equipment have to be located actually inside in the workshop near the production machinery where space is again at a premium. It must, therefore, be totally enclosed and so arranged that unauthorized personnel cannot interfere with it either to the detriment of production or to the danger of their own persons. In this connection the padlocking and interlocking of switches may be necessary and facilities should be provided for this.

Another matter which is attracting considerable attention at the present time is the fact that the larger organizations are requiring additional metering to obtain the necessary costing statistics for their administration. Proper facilities for mounting current-transformers and kilowatt-hour meters are being specified more and more frequently, hence the design of the gear should be such as to cater for this. Allied to this problem are those relating to the tariff requirements of the supply authorities. The tariff adopted can, incidentally influence the whole layout of the installation and should be considered from the very first stages.

One unfortunate modern tendency is that of load-shedding. An installation must, under present-day conditions, be so sectionalized that load-shedding is possible with the minimum of inconvenience. Usually this means that the essential loads must be segregated from the non-essential loads so that the latter can be reduced at will. A similar state of affairs exists where a kilowatt demand limit is included by the supply authority in their tariff agreement.

Power factor correction is also very much to the fore these days in order to satisfy tariff agreements and to make the maximum use of the existing copper. In some cases it is cheaper to install capacitors to take fuller advantage of existing cables than to lay down more copper. The distribution gear should be such as to cater for the connection of capacitors wherever it is decided to locate them.

The choice of equipment for the final distribution stages will be largely influenced by the type of factory in which it is installed. For normal process work which remains fairly static year after year the conventional distribution boards are usually preferred.

In industries like the motor car industry, however, where machine tools are constantly being repositioned, it is necessary to adopt a far more flexible method of connection. It is here that the overhead busbar system comes into its own and providing the design of busbar is such as to comply with the requirements already discussed, it is one of the most economical methods of distribution yet devised. Its use is, in fact, spreading to many other industries where previously its adoption had not been tried.

The overhead busbar system illustrated in Figs. 1.8, 1.13 and 1.14 has been specially designed to meet the requirements of those factories where the load is spread in units of varying sizes, such as individual drives, over the workshop area and where such units are liable to be moved to meet the varying demands of production.

Besides the claims for increased flexibility of distribution the advantages of overhead busbar in relation to other forms of distribution are that the labour and other costs of erection are small, saving can be effected on conduit and cables and there is little or no interference with the fabric of the building.

1.5 Other equipment for specific applications

The choice of equipment is, of course, very largely influenced by the conditions obtaining in the industry in which it is being used. It is impossible to mention all the

FIG. 1.8
Overhead busbar system in the engineering industry

possible requirements and circumstances which are met with in the various industries, but a few typical points from each of the following specific industries may serve to illustrate the diversity of conditions which are met with in practice.

1.5.1 The Engineering Industry

The majority of electrical loads in the engineering industry are machine-tool loads, most of which vary between 2–20 hp, but which may be as high as 50 hp or even more for special machines. Machine tools are easily moved and factories producing large numbers of similar components are frequently rearranged to meet the changes in the production programme. This is especially so in industries where annual changes of design are required to meet the markets. Conditions in this type of industry are comparatively clean except for swarf and cooling suds, but have a constant traffic feeding the lines and removing finished articles.

The best type of installation in the machine shops themselves is undoubtedly the overhead busbar system. The roof structures are usually well suited to support the busbars which, of course, give the necessary flexibility to cope with the most frequent of changes.

One of the main drawbacks, for instance, to underfloor cabling or ducting in a machine shop is the accumulation of swarf and soluble oil in the ducts, together with the fortnightly swilling of caustic soda which itself will corrode almost all

FIG. 1.9
*Dust, fumes and high ambient temperatures in the
steel industry*

FIG. 1.10
*Corrosive conditions in galvanizing shop
in steel wire drawing industry*

forms of cable protection. Any interference with the floor also makes the repositioning of machine lines difficult.

The diversity factors in the engineering industry can vary over wide limits due to the different types of load. Welding load is difficult from a distribution point of view, as are direct on-line compressor loads. P.f. can also be very low in some instances. On the whole, however, the atmospheric and other conditions are not particularly arduous and good-class equipment can be relied upon to give very many years' service.

1.5.2 The Steel Industry

The two main factors to be taken into account when choosing equipment for installation in the steel industry are the extremely rough usage and the corrosive and dust-laden atmospheres to which the gear will be subjected. The pace at which production is carried out, the very heavy weights which are handled, and the high ambient temperatures combine to make steelworks' duty one of the most arduous duties which distribution equipment is called upon to perform.

It is most essential in view of these conditions that the gear should be mechanically robust and that it should be rated for a continuous duty of twenty-four hours a day and seven days a week. The metal enclosures should be dustproof and substantial enough to withstand the corrosion due to sulphurous conditions. It is not unusual to find switchgear and fusegear covered in fine corrosive dust to a depth of a quarter of an inch or more, especially in the charging alleys. It

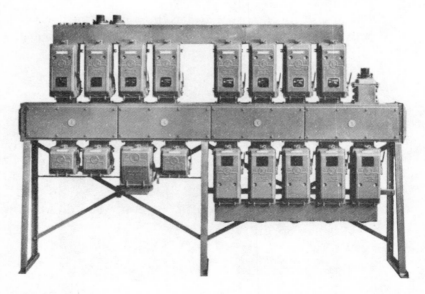

FIG. 1.11
Typical combination fuse switch/contactor board for
remote control in the chemical industry

will be appreciated that maintenance under these conditions is an extremely difficult matter and tends to become rather rough and ready because of the limited time which can be allowed for maintenance to be carried out. This is a further argument for the choice of generously rated equipment in the first place. It also points to the necessity for keeping the doors of all enclosures tightly closed to exclude the dust and justifies ideas to facilitate this, such as the cam-type door fasteners shown on the distribution board illustrated in Fig. 1.3.

One of the problems peculiar to the steel industry is that of controlling highly inductive loads, such as lifting magnets. Although these and similar devices are equipped with heavy-duty contactors for 'making' and 'breaking', it is inevitable that from time to time some person will break the load on the isolator or fuse switch. This may be due either to failure of the contactor or simply due to the whim of a crane driver. Hence it is important that all isolators or fuse switches associated with lifting magnets, etc., should be capable of successfully breaking the highly inductive load.

A considerable quantity of direct current is still used in the steel industry and this presents a few problems relative to the protective gear. Fortunately, the H.R.C. fuse and its associated equipment are quite capable of controlling and protecting d.c. circuits providing they are designed with this type of duty in mind. It is not unusual to find very high fault levels on the d.c. systems in steelworks because of the large generator capacities which may necessarily be connected on to one busbar. The protection of such systems is not always very easy, but many years of experience have shown that good-class fusegear equipment will successfully cope with the majority of faults which occur.

Not all locations in a steelworks are as bad as those mentioned and a considerable amount of equipment is used in places where standard industrial gear is quite satisfactory.

1.5.3 The Chemical Industry

Like the steel industry, the chemical industry is notoriously hard on distribution equipment. The hazards vary from corrosive liquids and vapours to solid salts and colloidal dusts on the one hand and highly inflammable substances on the other.

Heavy-duty equipment is the only class of equipment which will cope with these conditions in the majority of plants and even this has a comparatively short life on some.

Only constant maintenance and frequent painting can prolong the life of equipment which is subject to such severe usage, and a realistic attitude to the abnormal rate of depreciation is necessary on the part of managements to prevent strain on their maintenance engineers.

One of the methods which has been increasingly adopted in recent years to combat the adverse conditions in chemical plants, is that of siting the distribution and control gear in a clean position away from the production plant. The motors which may be totally enclosed or flameproof, as required, are necessarily fixed adjacent to the production plant although in some cases even these are situated in another room with the drive taken through a vapour-tight gland in the intervening wall. Where the motor is separated by some distance from its control gear, remote starting and stopping is usually provided by means of push-buttons near the motor. Occasionally the segregated switch rooms are pressurized at small pressure above atmosphere by means of a fan. Clean air from a reliable source is pushed into the room thus effectively stopping the ingress of pollution from the plant.

In the older plants where segregation has not been carried out, it is found that much of the corrosion takes place out of sight in inaccessible positions, such as the backs of wall-mounted distribution boards. The first sign of the effects of this kind of corrosion is usually an electrical breakdown inside the board due to the gradual disintegration of the back of the steel shell. It is, therefore, wise to avoid situations where this can arise either by providing access behind the board or taking extra precautions to protect the back. Otherwise it is prudent to carry a spare board or two to insure the production.

The materials used in distribution equipment for the worst situations in chemical plant have to be carefully chosen, but it is impracticable to stray too far from the

FIG. 1.12
Maintenance under difficulty in chemical industry

standard article. Good-class standard equipment of the heavy industrial pattern is usually the best proposition and, even when heavily depreciated, is more economical than equipment made to the special requirements and ideas of each individual concern.

1.5.4 The Textile Industry

The textile industry is still in the throes of a fairly rapid changeover from steam to electricity and the old line shafts are giving way to individual electric drives. Conditions, as they affect electrical equipment, are very diverse and vary between the clean conditions in the newer spinning and weaving sheds to the wet and steamy conditions of the dyeing trades. The rayon industry, which uses sulphuric and other acids in large quantities, has its own problems in combating the corrosive atmospheres.

The first phase in mill electrification was to introduce group drives to drive the existing line shafts. This gave greater control and increased the efficiency of production, but did not otherwise greatly alter conditions within the mills.

The second phase was to introduce individual drives on a large scale to increase both the quality and efficiency of production, to facilitate new methods of production and to improve conditions to attract labour. The elimination of line shafting meant that an extensive drive could be made to improve the appearance of weaving sheds and spinning rooms and all the service pipes and cables were arranged to be as unobtrusive as possible.

FIG. 1.13
Overhead busbar in a weaving shed

FIG. 1.14
An alternative method of mounting overhead busbar in the woollen industry

After a few years of this policy difficulties began to appear due to the inflexible nature of such installations. Whereas in the past looms and spinning frames, etc., were laid down and run in the same positions for half a century the modern tendency is for more and more flexibility. This varies in degree according to circumstances and is mainly due to the rapid changes which are being made in methods of production, as well as to developments of the actual machines and, to a large extent, to the reployment of labour.

Many firms have found it necessary to reposition looms and spinning frames several times in the course of the last ten years and this creates considerable difficulties affecting the electrical installation, especially when conduits and cables have been cemented into the floor. The answer to this problem is, of course, to use an overhead system but, at the same time, it has to be remembered that the circumstances which led to the adoption of underfloor systems are still valid. Any overhead busbar system which is adopted, while giving maximum flexibility, must not detract from the appearance of the factory. Figs. 1.13 and 1.14 illustrate how this has been achieved in actual installations.

The overhead busbar system also lends itself to easy extension and can easily cope with additional loads for ancillary services which are on the increase in all textile mills.

The lack of space for distribution equipment is particularly acute in some of the older mills where conversion has been necessary. Fire risk due to 'fly' is also a major factor to be considered and the electrical distribution system should be planned to reduce this hazard rather than increase it as can often happen if insufficient thought is given to the problem.

Conclusion

In a treatise of this nature it is inevitable that a great deal has to be left out. Thumbnail sketches of great industries cannot possibly tell the whole story, nor can the examples discussed apply to every case. This is not important so long as they serve to define the fundamentals which are the basis of a good distribution scheme and it is in this light that they should be regarded.

2 The property of non-deterioration in the 'English Electric' H.R.C. Fuse-link

The reinforcement of industrial and public supply systems in the post war decade due to rapid load growth brought into question the adequacy of existing equipment which had remained in service over many years with little or no opportunity for renewal. These questions were related to the problems of wear and tear and ageing rather than nominal rating and the question of possible deterioration of H.R.C. fuses was recognized as a possible problem. H.R.C. fuses had come into use in large numbers before and during the war and were now expected to cope with greatly increased loadings. The newly nationalized electrical supply industry would have faced a major and costly problem if it had found it necessary to renew fuses throughout the country. The research from which the paper was written proved that renewal was unnecessary. Existing fuses were able to remain in service with the full assurance that the safety and reliability of the system remained unimpaired.

'Non-deterioration' became established as a design parameter in fuse technology and had the effect of increasing the economic viability of fuse protection by several orders of magnitude. This capability has been kept under constant surveillance by checking fuses in service and in further research. The findings recorded in this paper have been amply substantiated by subsequent experience. Non-deterioration is not intrinsic in all H.R.C. fuses; it has to be achieved by stringent adherence to carefully defined design principles.

Originally published in the *'English Electric' Journal*, June 1954.

2.1 Introduction

One of the most important properties which has made the 'English Electric' cartridge fuse link a pre-eminent protective device, is that of non-deterioration. This is so because protective devices in service have to lie virtually dormant during most of their useful lives and to act only when a fault occurs. It has therefore become an established fact that the property of non-deterioration is of equal importance to any of the more active functions.

When the H.R.C. cartridge fuse-link was first introduced, it was designed to satisfy two important, and at that time urgent, requirements. The first was rupturing capacity, to cope with the increasing fault levels on supply systems, and the second was non-deterioration to overcome the serious disadvantages suffered by the types of semi-enclosed fuses in common use at that time. The semi-enclosed fuse had proved itself to be unreliable and unable safely to interrupt the growing values of short-circuit current occurring on supply networks and factory installations. The position was made even worse by the fact that this type of fuse, consisting essentially of a piece of bare wire suspended in free air, was susceptible to rapid oxidization and premature failure when carrying normal load currents. This gave rise to considerable abuse by encouraging users to strengthen the fuse wire, thus nullifying the effectiveness of the fuse as a protective device. Otherwise the interruptions in supply due to premature fuse blowing caused considerable inconvenience, and quite often loss of production. The fully-enclosed, cartridge-type fuse was born of necessity and has provided the solution to these and many other problems. Non-deterioration was and still is one of its most valuable attributes.

This was fully realized in the early stages of development of the 'English Electric' H.R.C. fuses some twenty-five years ago, when measures were adopted to ensure that the fuse would maintain constant characteristics throughout its working life. How well this policy was implemented has been amply illustrated by recent tests which have been carried out on cartridge fuses which have been in use for periods of up to twenty-five years. These tests are the culmination of other similar tests which have been carried out from time to time, and were made possible by the co-operation of the North Western Electricity Board of the British Electricity Authority who supplied fuse-links which had been in service in their own sub-stations. They were also able to supply data to substantiate the histories of the fuse-links with regard to loading and other service conditions.

The tests described in this publication are unique in that the English Electric Company is one of the very few concerns in the world which have been manufacturing fuse-links for a sufficiently long period to provide data for an actual life test of over a quarter of a century. Although enclosed fuses of various types had been in use early in the century, the modern conception of the high rupturing capacity cartridge fuse-link is a British development which was pioneered to a very large extent by this company.

The word 'non-deterioration' is one which has become identified by usage with H.R.C. fuse practice; it describes a characteristic property which has been proved in service by millions of 'English Electric' fuses since they were introduced over twenty-five years ago. The results of the tests now provide documented evidence that for all practical purposes a complete degree of non-deterioration has been achieved.

The question of non-deterioration is primarily one of preventing changes in the constituent materials of the fuse, but the problems involved in achieving the desired result cannot be so simply stated. Although the H.R.C. fuse is a static piece of equipment, it performs an active function in that it carries load currents of varying degrees throughout its life, and from time to time may have to withstand through-fault currents of considerable magnitude. It is subject to external influences which may vary from normal ambient conditions in clean atmospheres, to extremes

(b)
Close view of fuse assembly

FIG. 2.1(a)
Bromiley Cross Kiosk, installed 1927;

of temperature and atmospheric contamination as well as mechanical strain due to the vibration of associated equipment or adjacent machinery. In addition, it must be ready and capable at all times of safely interrupting short-circuit currents of all magnitudes up to the maximum fault level of the system in which it is installed. This latter condition provides the most stringent test of all.

When a fuse is interrupting a short-circuit current of a value normally to be expected under present-day conditions, it is controlling large values of potential power and is being subjected to a high degree of strain. It is under such strain that any changes of weaknesses will show themselves, and the smallest change from original condition may be magnified many times under fault conditions. Between the two extremes of normal load and fault conditions the fuses may have to handle

FIG. 2.2
Market Hall Substation, Bolton, installed 1934

many values of overcurrent due to overload and through-fault current, and will have to discriminate with other protective devices. For this purpose cartridge fuses are manufactured to conform to published time/current characteristic curves to which they must remain faithful throughout their working lives. Non-deterioration is thus related to every function which the fuse has to perform and every condition which it has to meet in service.

It will be seen that although the H.R.C. fuse as a protective device is simple in conception, its duties are both wide and complex, and its performance must be such as to cover satisfactorily the increasingly varied contingencies which arise in service.

Fuse design has undergone several changes since the H.R.C. cartridge fuse was first introduced. During the last decade particularly, fault conditions have increased in severity and fuse performance has had to be stepped-up to meet them. As a result, non-deterioration is more important than ever. An examination of the performance data of present-day fuses, which are described towards the end of this chapter will show that the tolerances allowable in all aspects of manufacture are being held to closer and closer limits. This applies not only to dimensional tolerances but also to the selection of grading of materials. The degree of non-deterioration must be of the highest order to ensure proper functioning of the fuse under the fault conditions which may occur years after it has left the factory.

2.2 Selection of Fuses for Non-deterioration Tests

The degree of non-deterioration in a particular design of fuse is difficult to assess because the only genuine trials are those of actual service. It is necessary to wait for a considerable number of years before being able to prove non-deterioration beyond all doubt. Such a period having passed since the 'English Electric' cartridge fuse was first introduced, steps were taken to carry out the appropriate tests.

A search was made for suitable fuses which were known to have been in service for the required period and whose histories could be vouched for accurately and with certainty. Of the many possibilities which presented themselves, it was decided to accept the generous offer of assistance from the North Western Electricity Board, on whose system were many fuses which were known to have remained in service unchanged for periods of eighteen to twenty-five years. A substantial number of fuse-links were taken from seven selected substations, each of which represented a different type of load, including heavy industrial, commercial, and domestic. Each of these substations had been under the direct supervision of the same engineer since they were first commissioned, and it was possible to obtain a personal guarantee from the engineer concerned that the fuses had remained undisturbed since they were first installed. In addition, it was possible to obtain a reasonable estimate of the ampere loadings to which the fuses had been subject both in winter and in summer during their years of service. Careful notes were also made of the ambient atmospheric conditions under which the fuses had been working.

TABLE 2.1
Severity of service

Fuse-link type and identity no.	Location	Year of installation of fuse-link	Details of loading (A)
D200F	Edgeworth Kiosk,		(Summer)
ND100	no. 1 circuit	1927	50
ND101			(Winter)
ND102			100
D200F	Edgeworth Kiosk,		(Summer)
ND103	no. 2 circuit	1927	50
ND104			(Winter)
ND105			100
D200F	Bromiley Cross		(Summer)
ND106	Kiosk	1927	50
ND107			(Winter)
ND108			100

D200F ND109	Harwood Kiosk	1927	(Summer) 50 (Winter) 100
28T250F ND110 ND111	Market Hall Substation, no. 3 circuit	1934	(Summer) 125 (Winter) 210
28TK300 ND112	Market Hall Substation, no. 3 circuit	1934	(Summer) 150 (Winter) 250
29T300K ND113 ND114 ND115	Market Hall Substation, no. 1 circuit	1934	(Summer) 150 (Winter 250
30T350M ND116 ND117 ND118	Market Hall Substation, no. 2 circuit	1934	(Summer) 175 (Winter) 300
TM400 ND119 ND120 ND121	Weaste Road South Substation no. 1 feeder	1935	(1935) 100 (1939) 230 (1952) 250
TM400 ND122 ND123 ND124	Weaste Road South Substation, no. 2 feeder	1935	(1935) 120 (1939 to 1952) increase to 300
TM400 ND125 ND126 ND127	Gore Avenue Substation, Weaste Lane West Side	1935	(1935) 50 (1942) 150 (1952) 200
TM400 ND128 ND129 ND130	Gore Avenue Substation, Weaste Lane East Side	1935	(1935) 50 (1942) 150 (1952) 200
TKF300 ND131 ND132 ND133	Bolton Street Substation	1936	(Summer) 150 (Winter) 250

Note: All substations subject to peak winter overloads.

FIG. 2.3
Weaste Road South Substation, Salford, installed 1935

The actual age of many of the fuse-links was further checked by reference to the inspection marks which were put on the component parts of the cartridges during manufacure, and in some cases by noting small features of design which were known to be contemporary with the types of fuse manufactured at the time the selected fuse-links were installed. By these means the accuracy of all data and periods of duty were checked beyond all reasonable doubt.

Table 2.1 gives a list of the fuse-links selected, with brief details of site location, year of installation and loadings. The identification numbers allocated to the fuse-links are used throughout the subsequent illustrations.

Figs. 2.1, 2.2 and 2.3 show interior views of some of the substations from which the selected fuse-links were taken, and indicate in some measure the conditions under which they worked in service.

The steel kiosk-type substation shown in Fig. 2.1(a) serves a domestic and rural network and was commissioned prior to 1927. A close view of one of the fuse handles and contacts is shown in Fig. 2.1(b), and it will be seen that this equipment has acquitted itself well, remaining in excellent condition over a period of twenty-five years and showing promise of good service for many years to come.

The fuse-links selected from this location were of 150 A rating (in accordance with B.S. 88: 1952), and were of a type which was so old as to have become obsolete in general design almost twenty-five years ago. They nevertheless embodied all the features relevant to the study of non-deterioration. The fuses had carried loads varying between 50 and 100 A with peaks of higher value during winter conditions. It was difficult in this case to obtain accurate data concerning the actual value of the highest winter loads, but it is estimated that these would have exceeded 150 A during extremely cold periods.

The substation fuseboard shown in Fig. 2.2 is of very old design which had been specially adapted to accommodate the then newly introduced cartridge fuses. The ambient temperatures in this substation were fairly high due to a consistently high town load, and certain difficulties in substation ventilation. The fuses selected for test were of both 300 and 350 A capacity and for eighteen years had carried summer loads of 150 A and 175 A respectively, and winter loads of 250 A and 300 A respectively, both with peaks of overload during very cold or severe weather.

The substation fuseboard shown in Fig. 2.3 is of the 'English Electric' Skeltag design serving a mixed industrial and domestic load and subject to a considerable degree of atmospheric pollution. The fuses selected in this case were of 400 A capacity and had carried loads increasing from 100 A in 1935 to 250-300 A in 1952, again with winter overloads, which it is estimated would sometimes have exceeded the nominal rating of the fuses for short periods. It is also considered inevitable that all the selected fuse-links would at some time or other have carried through-fault currents due to faults on branch networks or on consumers' installations.

2.3 Investigation to Determine Physical and Electrical Condition of the Selected Fuse-Links

The 'English Electric' fuse-link consists principally of a ceramic body, pure silver elements, clean silica quartz, asbestos washers, porcelain plugs, brass end-caps and copper tags, the last two items being electrotinned. The assembly also includes solder of various types, cement and indicator devices. Deterioration must involve a change in one or more of these materials or a change in their structure, and the tests described were devised to explore every possible aspect of the factors involved.

A total of thirty-four fuse-links was available for investigation, and as some of the tests and examination contemplated involved the blowing or dismantling of some of the samples at various stages, the programme was arranged to ensure that a full complement of tests was carried out on each type of cartridge.

In order to provide a basis of comparison in each of the tests, a number of fuse-links were made up exactly to the specification which obtained when each type of those under test was originally manufactured, and great care was taken to use the identical materials originally used and to follow the original methods of manufacture. The programme was then so arranged that the old and new fuses could be tested simultaneously whenever possible. This enabled the comparisons obtained to be checked and studied at each stage.

The testing programme resolved itself into the investigation of electrical performance, preceded by resistance measurements and a thorough physical examination.

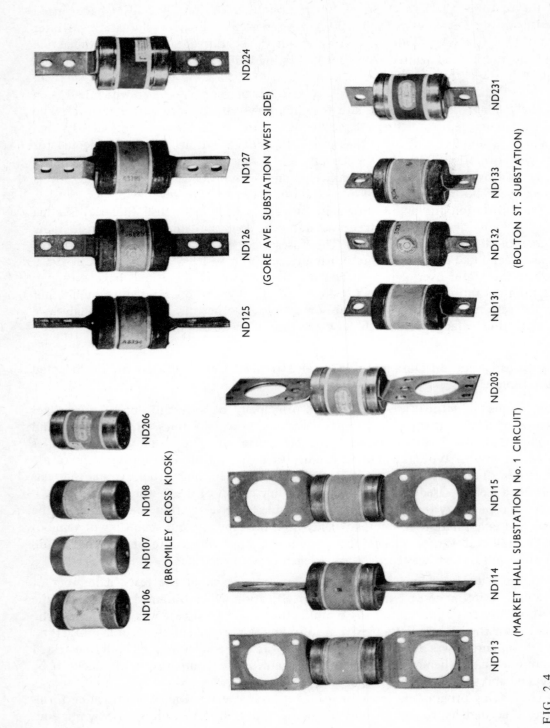

ND224

ND231

ND127   ND126   ND125

(GORE AVE. SUBSTATION WEST SIDE)

ND133   ND132   ND131

(BOLTON ST. SUBSTATION)

ND206   ND108   ND107   ND106

(BROMILEY CROSS KIOSK)

ND203   ND115   ND114   ND113

(MARKET HALL SUBSTATION No. 1 CIRCUIT)

FIG. 2.4
*Fuse links as taken from service and compared with new fuse links of similar
construction. Scale one-quarter full size*

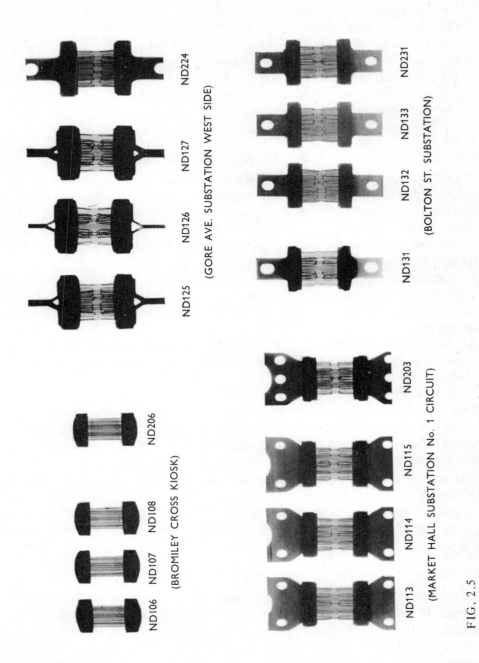

ND224 ND127 ND126 ND125

(GORE AVE. SUBSTATION WEST SIDE)

ND231 ND133 ND132 ND131

(BOLTON ST. SUBSTATION)

ND206 ND108 ND107 ND106

(BROMILEY CROSS KIOSK)

ND203 ND115 ND114 ND113

(MARKET HALL SUBSTATION No. 1 CIRCUIT)

FIG. 2.5
Radiographs of the fuse links shown in Fig. 2.4

35

These may be arbitrarily listed as follows:

- (a) visual exterior examination;
- (b) radiograph examination;
- (c) visual interior examination, including examination of each component part;
- (d) x-ray diffraction tests of constituent materials;
- (e) measurement of fuse-link resistance;
- (f) temperature rise tests at full rated current;
- (g) determination of minimum fusing currents;
- (h) short-circuit tests at various critical values of prospective short-circuit current.

2.3.1 *Visual Condition of Selected Fuse-Links*

The external conditions of the selected fuse-links, some of which are shown in Fig. 2.4, indicates the effect of atmospheric conditions on the fuse caps, tags and ceramic bodies. The figure shows the selected fuses compared with similar fuses of new construction.

It will be seen that the contact surface on the fuse tags has been maintained in good condition in all the examples.

In contrast to the contact surfaces, the end-caps of the selected fuse-links show various degrees of oxidization and surface contamination, which in each case is consistent with the atmospheres and ambient conditions in which the fuses have been working.

All soldered joints between tags and end-caps are good, the ceramic bodies are unimpaired, and the cement joints are as effective as when the fuses were new.

2.3.2 *Examination by Radiography*

All the selected fuse-links were radiographed, and Fig. 2.5 shows the X-ray plates corresponding to the fuse-links illustrated in Fig. 2.4. As before, the plates are arranged in sets to show a comparison between the old and new fuse-links of similar construction. The undisturbed interior condition of the fuse-links can be quite clearly seen. By careful interpretation it is possible to check that the gap portion of each element is adequately spaced from the others, and it is in every case properly positioned at the centre of the fuse-link body. It will also be seen that there has been no thinning of the elements due to corrosion, or sagging due to mechanical or thermal stress.

2.3.3 *Visual Condition of Interior Components*

Fig. 2.6 shows two examples from the selected fuse-links with the outer end-caps removed. It will be seen that the outer end-caps are solidly soldered to the inner end-caps, and the silver elements are in turn each soldered to the inner end-cap. It is considered that dry joints should be permitted only in positions where they can be checked periodically or when large enough and solid enough to be beyond doubt. The internal connections of the fuse-link do not come within this category, as they

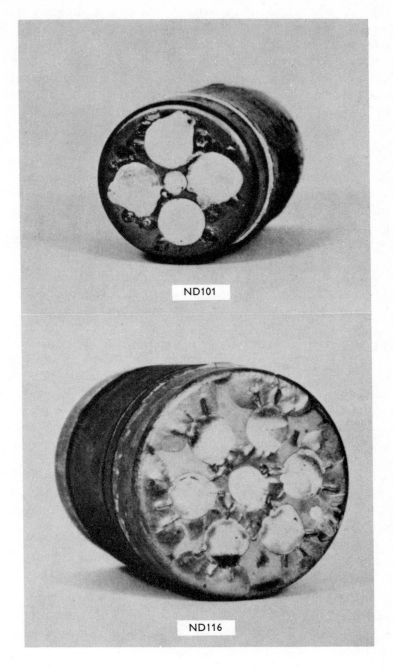

ND101

ND116

FIG. 2.6
*End views of two samples of fuse-links, with outer caps removed,
showing condition of sealing, plating and soldering*

are hidden out of sight and are therefore out of mind. Fig. 2.6 shows the conditions of the soldered joints and is evidence of how this design feature has been implemented with success.

The advantage of soldering the elements to the inner end-caps is that the solder makes a perfect electrical contact without affecting the structure of the materials being joined together. Other methods of making the connection, such as welding, tend to reduce the cross section of the elements and to distort the surface to which the elements are joined. It is for this reason that the welding of elements (by methods at present available) is to be deprecated.

The hermetic sealing of the elements within the cartridge is seen in Fig. 2.6 to have remained completely effective. The cement seals show no sign of disintegration, a remarkable achievement when it is considered that the fuse-link, as a thermal device, has been subject to innumerable temperature cycles, involving expansions and contractions of the end-caps and cement alike, over a long period of years.

The internal components of the same cartridges are shown in Fig. 2.7. The silver elements are seen to be as bright as in their pristine condition. The ceramic plugs and asbestos washers are in good condition and the quartz filler is completely un-changed. A test for moisture content of the internal parts also showed that there had been no significant absorption of moisture by any of the material concerned.

The examination of the fuse elements also included a photomicrograph examina-tion. Fig. 2.8 shows a series of plates which again provide comparisons between the old and the new elements. It will be seen from both the sectional and longitudinal profiles that there is no evidence of surface erosion. The plates showing the sections of soldered joints also prove that no change has taken place in the contact between the solder and the silver elements. A true soldered joint should give completely intimate contact without affecting the surface of the metal to which the solder is attached. This result has been successfully achieved on each of the samples submitted for examination, and no change could be detected in the effectiveness of any of the joints. It can be seen that the mixture of solder has been varied slightly between the old and the new elements, but this has no significant bearing on the factors affecting deterioration. For those to whom this method of presentation is not familiar it may be useful to explain that the horizontal wire shown on each of the soldered joints is actually a transverse section through a curved circular wire. This gives the appearance of narrowing towards the middle, but it will be appreciated that this is a function of the method by which the wire has been cut and that the actual wire has not been reduced in section.

2.3.4 X-ray Diffraction Tests

Another method which was tried in an effort to detect changes in the fuse elements was that involving X-ray diffraction tests, which were carried out on an experimental basis in the spirit of 'leaving no stone unturned'. It was not expected that conclusive evidence could be obtained, as it is obvious that the only true basis of comparison would have been successive tests taken on the same pieces of material at the beginning and the end of the period under review. Although such a

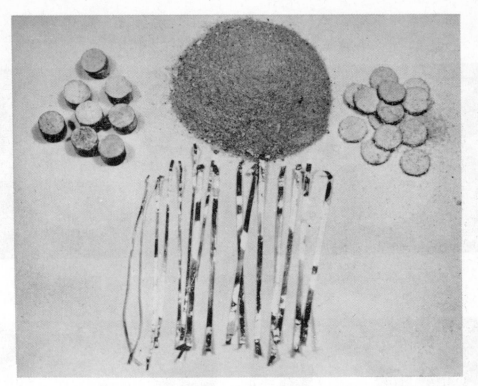

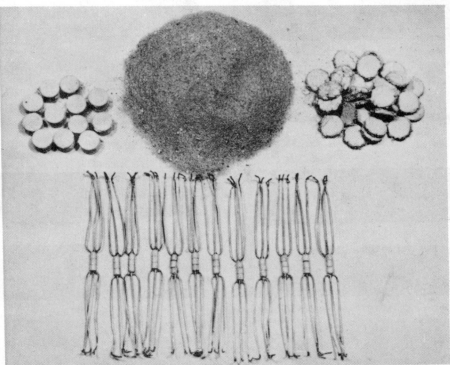

FIG. 2.7
Internal components of fuse-links shown in Fig. 2.6 comprising elements, filler, sealing washers and ceramic plugs

(a)   TRANSVERSE SECTION OF GAP PORTION,   (b)
MAGNIFICATION x 77

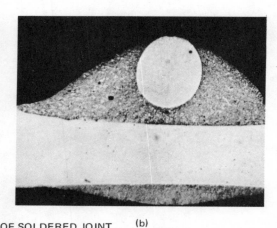

(a)   TRANSVERSE SECTION OF SOLDERED JOINT,   (b)
MAGNIFICATION x 35

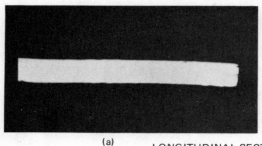

(a)   LONGITUDINAL SECTION OF GAP PORTION,   (b)
MAGNIFICATION x 8½

FIG. 2.8
Photomicrographs showing sectional and longitudinal profiles of silver elements: (a) sample taken from fuse-link no. ND115, installed 1934; (b) sample taken from new fuse-link of similar construction

40

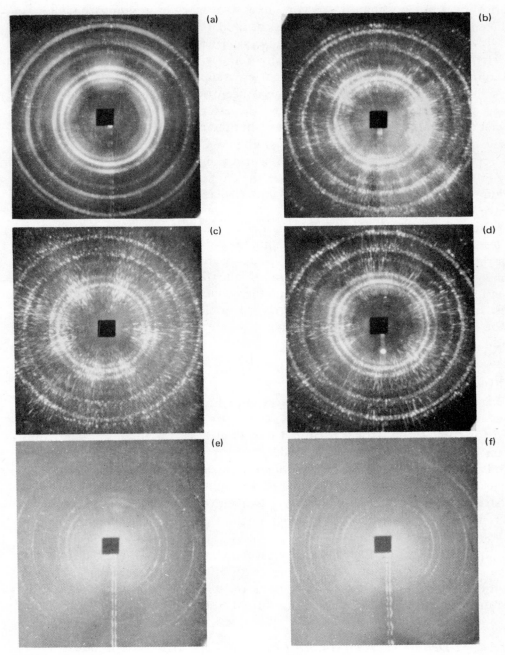

FIG. 2.9

Photographic records of X-ray diffraction tests on silver fuse elements: (a) 0.002 in x 0.04 in silver strip element, of new construction; (b) 0.002 in x 0.04 in silver strip element, after 25 years service; (c) 0.002 in x 0.07 in silver strip element, of new construction; (d) 0.002 in x 0.07 in silver strip element after 25 years' service; (e) type 30 silver wire element, 0.022 in x 0.022 in, of new construction; (f) type 30 silver wire element, 0.022 in x 0.022 in, after 18 years' service

41

comparison was not possible, it was considered worth while to compare samples of silver taken from the old and new fuse-links of similar construction.

The investigation failed to detect any significant change which might be interpreted as deterioration. The plates shown in Fig. 2.9 are typical of some of the results obtained, and the characteristic patterns appearing may be interpreted as follows: the number and disposition of the concentric rings indicate the crystalline type and structure of the material (silver), the asterism appearing in the form of radial marks indicates the degree of strain within the material, and the continuity or otherwise of the concentric rings indicates grain size. Continuous and unbroken rings appear for fine grain, and broken rings for coarse grain. In Fig. 2.9, (a) and (b) show slightly greater strain and larger grain size in the old sample than in the new, but this condition seems to be reversed in (c) and (d); (e) and (f) show no apparent difference between the old and new samples.

These tests are not entirely conclusive when considered out of context with the other tests of a more practical nature, but they do show that slight degrees of interior stress and variations in grain size are not significant to the proper performances of the fuse. This is shown by the fact that the worst condition detected in the old samples is no worse than the worst condition detected in the new samples. In other words, it is clear that the silver elements of the older fuse-links have not suffered change beyond the point where the electrical conductivity is affected.

2.3.5 Measurements of Fuse-Link Resistance

Table 2.2 shows the values of resistance (in microhms) recorded for a number of the fuses under test. The values for similar fuse-links of new construction are also included for comparison. It will be found that in all cases the fuse-links fall within the permitted tolerence of ±5%. Improved methods of quality control in present-day production make it possible to obtain a higher degree of consistency in resistance values than is indicated by some of the figures in the table. While this fact is not strictly relevant to the objective investigation of non-deterioration, it is necessary to make it clear in order to avoid misconceptions concerning present-day values.

Apart from the fact that the resistance measurements prove that the old fuses are still well within the permitted tolerances allowed under present-day standard specifications, they are necessary as a basis on which to judge the results of temperature rise tests and minimum fusing current tests.

2.3.6 Tests to Determine Temperature Rise Values at Rated Current and to Find the Minimum Fusing Current of the Fuse-Links

The maximum temperature rise for various parts of the fuse assembly and the minimum fusing current expressed as a fusing factor are defined for present-day designs of fuse-link in B.S. 88:1952. The methods used to comply with these

TABLE 2.2
Measurements of resistance

Fuse-links of 16 to 25 years' service			Similar fuse-links of new construction		
Fuse-link identity no.	Resistance $(\mu\Omega)$	Percentage Variation from designed value	Fuse-link identity no.	Resistance $(\mu\Omega)$	Percentage Variation from designed value
ND100	930	+1.1	ND200	910	−1.1
ND101	925	+0.5	ND201	930	+1.1
ND102	930	+1.1	ND202	920	0.0
ND103	925	+0.5	ND203	935	+1.6
ND106	920	0.0	ND206	900	−2.2
ND104	910	−1.1	ND204	910	−1.1
ND107	920	0.0			
ND108	960	+4.4			
ND105	930	+1.1			
ND109	910	−1.1			
ND110	214	+4.4	ND210	200	−2.4
ND111	212	+3.4	ND211	197	−3.9
ND112	190	+3.2	ND212	184	0.0
			ND212A	183	−0.55
			ND212B	182	−1.1
ND113	205	−3.7	ND213	219	+2.8
ND114	209	−1.9	ND214	220	+3.3
ND115	208	−2.3	ND215	220	+3.3
ND116	182	+1.1	ND216	176	−2.2
ND117	182	+1.1	ND217	175	−2.8
ND118	182	+1.1	ND218	173	−3.9
ND119	147	+0.7	ND219	143	−2.0
ND120	141	−3.4	ND220	143	−2.0
ND121	144	−1.4	ND221	145	−0.7
ND122	142	−2.7	ND222	143	−2.0
ND123	140	−4.1	ND223	145	−0.7
ND124	153	+4.8	ND224	148	+1.4
ND125	148	+1.37			
ND126	149	+2.0			
ND127	146	0.0			
ND128	148	+1.4			
ND129	151	+3.4			
ND130	141	−3.4			
ND131	192	+4.3	ND231	188	+2.2
ND132	179	−2.7	ND232	185	+0.55
ND133	180	−2.2	ND233	183	−0.55

FIG. 2.10
Test rig, showing method of mounting and connecting fuses for temperature rise
tests

requirements were followed in this investigation. Fig. 2.10 shows the apparatus
which was used to obtain the results tabulated in Table 2.3. It can be seen that the
comparisons between the old fuse-links and the new fuse-links of similar construction are well within the permitted tolerances. Such small divergencies as do occur
are not incompatible with the physical characteristics of the individual fuse-links;
for instance, a slightly lower resistance gives a slightly higher value of minimum
fusing current.

The method of carrying out the temperature rise tests is to apply the full rated
current to the fuse and to record the actual temperatures of the top fuse contact,
top fixed terminal, and a position on the connecting cable approximately 12 in
from the top fuse contact. Readings are taken every 15 min and continued until
such time as there is no appreciable difference in temperature over three successive
readings. Fig. 2.11 gives in graphical form a typical example of the results of such a
test.

The determination of the minimum fusing current is a continuation of the
previous test, in which the current is increased by regular steps of 10 per cent the

TABLE 2.3
Examples of heating and fusing tests

Fuse-link no.	ND 100	ND 200	ND 131	ND 231	ND 124	ND 224
Year installed	1927	New fuse-link of similar construction	1936	New fuse-link of similar construction	1935	New fuse-link of similar construction
Fuse-link resistance ($\mu\Omega$)	930	910	192	188	153	148
Complete fuse resistance ($\mu\Omega$)	1,250	1,200	250	245	248	230
Test current (A)	150	150	300	300	400	400
Temperature rise (°C) at rated current						
Top fixed terminal*	34	34	27	26	31	29
Top fuse terminal†	47	50	37	34	40	36
Ambient temperature (°C)	20	20	25	25	28‡	28‡
Minimum fusing current (A)	220	220	420	450	560	600

*Permitted maximum 46°C in accordance with B.S. 88:1952.
†Permitted maximum 55°C in accordance with B.S. 88:1952.
‡The high ambient temperatures are due to the fact that the tests were carried out on a very hot day in summer.

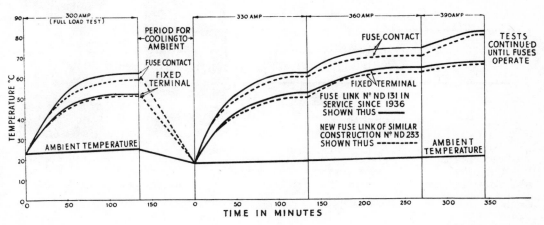

FIG. 2.11
Curves showing comparison between temperature rise characteristics of old and new fuse-links of similar construction

temperature readings being allowed to stabilise for each step before proceeding to the next. This process is continued until the fuse link blows. Fig. 2.11 also includes a typical example illustrating the methods used.

2.3.7 Short-circuit Tests

As mentioned previously, the short-circuit test was regarded as being the most critical of all. Other electrical tests were mainly concerned with the ability of the fuse to carry load under normal conditions. Under such conditions if deterioration had occurred it would usually result in premature blowing, and although this could

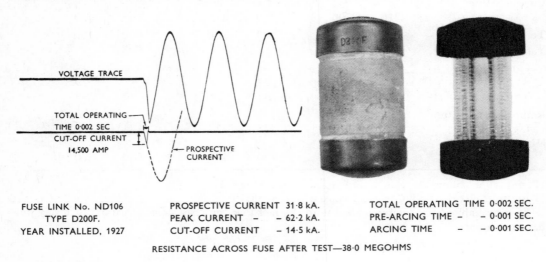

FUSE LINK No. ND106	PROSPECTIVE CURRENT 31·8 kA.	TOTAL OPERATING TIME 0·002 SEC.
TYPE D200F.	PEAK CURRENT – – 62·2 kA.	PRE-ARCING TIME – – 0·001 SEC.
YEAR INSTALLED, 1927	CUT-OFF CURRENT – 14·5 kA.	ARCING TIME – – 0·001 SEC.

RESISTANCE ACROSS FUSE AFTER TEST—38·0 MEGOHMS

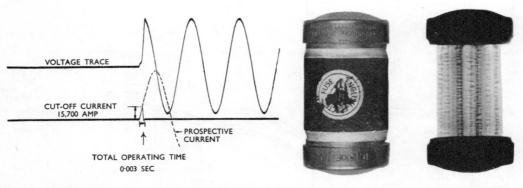

FUSE LINK No. ND256	PROSPECTIVE CURRENT 32·2 kA.	TOTAL OPERATING TIME 0·003 SEC.
TYPE D200F.	PEAK CURRENT – – 62·2 kA.	PRE-ARCING TIME – – 0·001 SEC.
YEAR INSTALLED, NEW	CUT-OFF CURRENT – 15·7 kA.	ARCING TIME – – 0·002 SEC.

RESISTANCE ACROSS FUSE AFTER TEST—3·0 MEGOHMS

FIG. 2.12(a)
Short-circuit tests at 33 kA to compare cut-off characteristics of old and new fuses of similar construction

cause serious inconvenience it would not involve hazards to personnel or equipment. It is interesting to note that deterioration in a fuse would normally cause it to 'fail to safety' under overload conditions.

The short-circuit test, on the other hand, imposes the maximum strain on the fuse-link both electrically and mechanically, and the criterion of success is that the fuse-link should interrupt the short-circuit current without bursting or in any way damaging associated equipment.

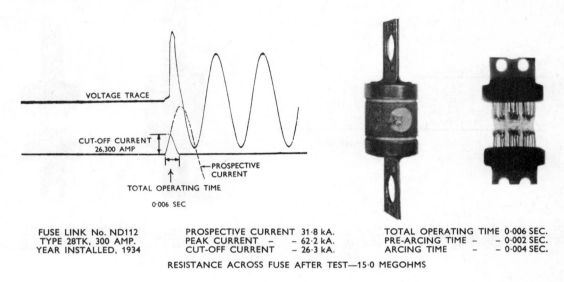

FUSE LINK No. ND112   PROSPECTIVE CURRENT 31·8 kA.   TOTAL OPERATING TIME 0·006 SEC.
TYPE 28TK, 300 AMP.   PEAK CURRENT –   – 62·2 kA.   PRE-ARCING TIME –   – 0·002 SEC.
YEAR INSTALLED, 1934   CUT-OFF CURRENT   – 26·3 kA.   ARCING TIME   –   – 0·004 SEC.

RESISTANCE ACROSS FUSE AFTER TEST—15·0 MEGOHMS

FUSE LINK No. ND257   PROSPECTIVE CURRENT 32·1 kA.   TOTAL OPERATING TIME 0·005 SEC.
TYPE TKF, 300 AMP.   PEAK CURRENT –   – 62·2 kA.   PRE-ARCING TIME –   – 0·002 SEC.
YEAR INSTALLED, NEW   CUT-OFF CURRENT   – 26·0 kA.   ARCING TIME   –   – 0·003 SEC.

RESISTANCE ACROSS FUSE AFTER TEST—100·0 MEGOHMS

FIG. 2.12(b)
Short-circuit tests at 33 kA to compare cut-off characteristics of old and new fuses of similar construction

When an H.R.C. fuse interrupts a heavy fault it exhibits an ability to limit the short-circuit current. This ability is referred to as a 'cut-off' and has the effect of reducing the magnetic and thermal stresses both in the system and within the fuse itself under fault conditions. Cut-off is in fact one of the main reasons why the H.R.C. fuse is so successful as a protective device.

In deciding the tests which should be carried out for non-deterioration it was

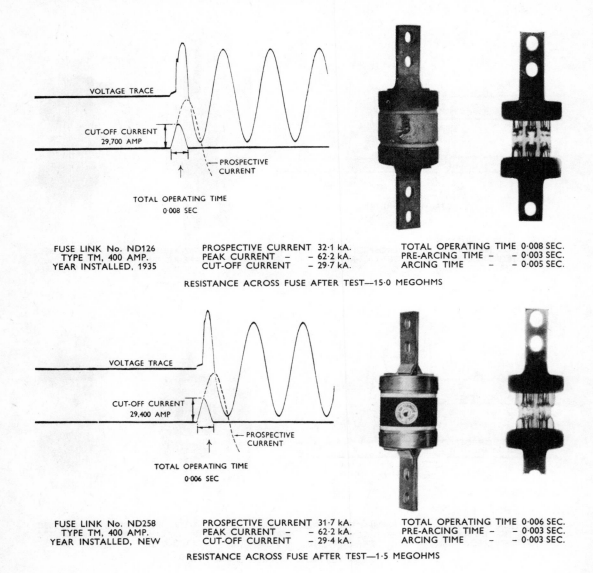

FIG. 2.12(c)
Short-circuit tests at 33 kA to compare cut off characteristics of old and new fuses of similar construction

agreed that a study of cut-off characteristics would provide the most fruitful and critical proofs of possible changes in short-circuit performance due to long service.

The oscillograms shown in Fig. 2.12 (a), (b), (c) and (d) illustrate results of some of the tests. The object of the tests, which were carried out in the Nelson Research Laboratories at Stafford, was to obtain values of cut-off currents and operating times with which to draw an accurate comparison between the old and new fuses of

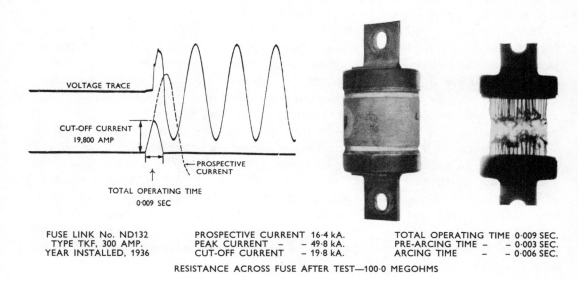

FUSE LINK No. ND132   PROSPECTIVE CURRENT 16·4 kA.   TOTAL OPERATING TIME 0·009 SEC.
TYPE TKF, 300 AMP.   PEAK CURRENT – – 49·8 kA.   PRE-ARCING TIME – – 0·003 SEC.
YEAR INSTALLED, 1936   CUT-OFF CURRENT – 19·8 kA.   ARCING TIME – – 0·006 SEC.

RESISTANCE ACROSS FUSE AFTER TEST—100·0 MEGOHMS

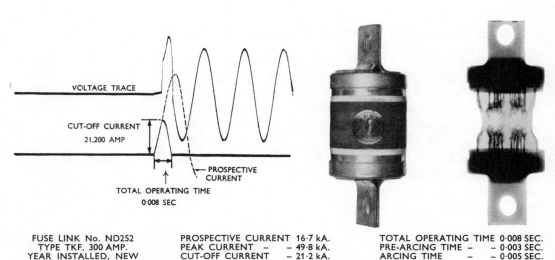

FUSE LINK No. ND252   PROSPECTIVE CURRENT 16·7 kA.   TOTAL OPERATING TIME 0·008 SEC.
TYPE TKF, 300 AMP.   PEAK CURRENT – – 49·8 kA.   PRE-ARCING TIME – – 0·003 SEC.
YEAR INSTALLED, NEW   CUT-OFF CURRENT – 21·2 kA.   ARCING TIME – – 0·005 SEC.

RESISTANCE ACROSS FUSE AFTER TEST—100·0 MEGOHMS

FIG. 2.12(d)
Short-circuit tests at 16.5 kA to compare cut-off characteristics of old and new fuses of similar construction

similar construction. It will be seen from the oscillograms that this was achieved by careful control of the test circuit conditions (which were exactly similar to those laid down in B.S. 88:1952) and by switching on to the fault at a pre-determined point on the voltage wave in each case.

The resulting current traces obtained show a remarkable degree of similarity between the old and the new fuses. Not only is there agreement between the respective peak values and operating times, but the shapes of both the pre-arcing and arcing portion of the traces are similar in characteristic.

It will be seen that two values of prospective short-circuit current were chosen, i.e. 33 000 A and 16 000 A, both values being symmetrical r.m.s. and corresponding to British Standard categories AC4 and AC3 respectively. These values are representative of the most onerous conditions which are likely to have been encountered in service. Typical oscillograms and test data from the actual test reports are shown in Figs. 2.12(a), (b), (c) and (d), together with photographs and radiographs showing the condition of the fuse-links after successfully interrupting the short-circuit. It is interesting to observe the values of arc voltage shown in the oscillograms. In no case do these exceed the system voltage by more than a small margin, thus illustrating another advantage which is inherent in 'English Electric' H.R.C. fuse protection. The resistance measurements of the blown fuses taken immediately after test leave no doubt that the circuit has been effectively interrupted. The tests were entirely conclusive and in no case was there any evidence of deterioration in respect of the ability of the old fuses to operate under heavy fault conditions.

2.4 Non-deterioration in Relation to Present-Day Practice

The real value of the foregoing tests is twofold. Firstly, they justify the methods which have been adopted for achieving the highest possible degree of non-deterioration, and confirm objectively what has already been proved in service throughout the world. Secondly, they provide guidance for directing future design policy. New developments in fuse design and new methods of production are constantly being investigated, and every decision concerning them must be ruled by the necessity for maintaining and if possible improving upon established standards.

Non-deterioration is a property which becomes more important as the performance of the fuse is increased. It has been shown that the old fuse-links described in the foregoing tests have been able to cope successfully with short-circuit current up to 33 000 A (r.m.s. symmetrical) at normal system voltages. Since they were manufactured, however, the increase in fault levels has called for fuses of higher short-circuit performance. In the case of British Standard 88, the 1947 edition introduced category AC5 to deal with short-circuit currents up to 46 000 A r.m.s. symmetrical, and the Company's range of 'T'-type fuses has received certificates of rating by the Association of Short Circuit Testing Authorities for this duty. In Canada a new specification C22.2 No. 106 covering H.R.C. fuses calls for still higher rupturing capacity, namely, 100 000 amperes r.m.s. asymmetrical (corresponding to 80 000 A r.m.s. symmetrical) at 600 V a.c. The company has designed a new range of fuses, type 'C', which have been successfully tested for this duty and have received Approval No. 12203 from the Canadian Standards Association.

The tests for non-deterioration also provided valuable evidence in another direction. All specifications and proving tests presuppose that the manufacturer will produce equipment for general sale which is faithful in every respect to declared characteristics and identical with the samples which have successfully passed the prescribed tests. This is especially important in the case of the cartridge fuse-link, because its functioning parts are totally enclosed and cannot be inspected. Its success as a protective device depends entirely upon its soundness of design and the skill by which it is manufactured. In view of this the reputable manufacturer is obliged to adopt techniques which leave no doubt that each fuse produced will perform in accordance with accepted specifications.

The question of consistency in manufacture and performance is constantly under review in this company, and in addition to taking resistance measurements of every fuse manufactured, tests are continually in progress to check consistency under normal service and fault conditions. The result of this is that a high degree of consistency has been achieved on every fuse and range of fuses manufactured. The tests for non-deterioration now prove that this consistency can in fact be maintained throughout the life of the fuse.

Although the samples chosen for test were taken from a public supply network, the results obtained apply equally to other applications. H.R.C. fuses are used extensively in industrial, marine, traction, and aircraft installations, and the requirements of non-deterioration are equally valid to them all.

3 Discrimination between H.R.C. Fuses

The 1950s saw a vast increase in usage of electrical energy and in the size and load density of industrial installations. In power stations and process industries automatic control brought greater system complexity. These conditions gave rise to a demand for more accurate performance and predictability of protective equipment. Greater emphasis was put on the fidelity of fuses in relation to their published characteristics so that selectivity and discrimination could be improved. Many large installations required interconnection of high-voltage supplies which in turn had to be co-ordinated with the low-voltage circuits. Thus the order of selectivity and discrimination which had been required in high-voltage systems and which had been developed to an advanced degree was extended into low-voltage installations.

Fuse designers responded to these demands by producing fuses of greater accuracy. Discrimination becomes critical at the higher levels of short-circuit current and it is in this zone that accuracy becomes important. The principles upon which discrimination can be defined and specified with certainty required extensive short-circuit testing at full power to provide empirical proof. This was carried out and the basis for fuse discrimination duly established. The work was subsequently reflected in British and other Standards.

Originally published in January 1959 by The Institution of Electrical Engineers.

3.0 Terms and definitions

Definitions are generally as laid down in B.S. 88: 1952 and as illustrated in Fig. 3.1.

The term 'H.R.C. fuse' may be taken to refer to those fuses having rupturing capacities of the order of AC4, DC4 (i.e. 33 kA r.m.s. symmetrical) or above. The abbreviation H.R.C. (high rupturing capacity) is used because of its wide acceptance over many years, although by implication in B.S. 88: 1952 the term 'high breaking capacity' is preferred.

3.1 Introduction

Discrimination between H.R.C. fuses has been achieved with practical success since these devices were first introduced. Progress in this respect has been mainly empirical and has been taken for granted by the majority of users. During the last decade the increased use of H.R.C. fuses has given rise to an increased interest in the theory of fuse operation. New knowledge has emerged to confirm or modify previous theories in the light of experience, and a wider appreciation of their capabilities has opened up new opportunities of extending their usefulness.

The purpose of this chapter is twofold: first, to define the capabilities and limits of the modern H.R.C. fuse as regards discriminative protection and to indicate the extent of its further usefulness; secondly, to relate the theory of fuse performance to standard distribution practice. Although the treatment of the subject is kept within practical limits in relation to standard practice, it is recognized that special applications occur which require more detailed treatment and where the approximations permitted in standard practice would not be appropriate. The principles on which discrimination depend, however, are the same for all cases.

The subject is topical in as much as many users who have had practical experience of fuses over many years are now taking an interest in the more technological aspects of fuse performance and are interested in knowing how the new knowledge can be reconciled with established practice.

The keynote in the achievement of the H.R.C. fuse as regards discrimination is its basic simplicity of construction, which permits consistency in manufacture and accuracy of characteristics. This does not mean that there is no subtlety in design, nor does it imply that the duty of the fuse is simple — on the contrary, fuse operation involves all the technology implied in circuit interruption. The theory of fuse discrimination must ultimately be based upon an appreciation of fuse operation under various short-circuit conditions. It must be realized that the principal duty of the fuse is to interrupt short-circuits and that discrimination is usually regarded as a secondary requirement. Nevertheless, a satisfactory compromise between the two can be achieved.

From a consideration of the technological factors, discrimination depends on the correct application of the fuse in any given situation. Practical and economic considerations external to the fuse itself can sometimes significantly affect the results obtained.

It should be emphasized that the remarks and opinions expressed can be taken to apply only to the modern H.R.C. fuse manufactured to high standards.

3.2 Object of discrimination and the basic functions involved

3.2.1 Definition

Discrimination may be defined as the ability of protective devices to interrupt the supply to a faulty circuit without interfering in any way with the source of supply or the remaining healthy circuits fed from it. This necessarily implies co-ordination between two interdependent devices in series.

3.2.2 System requirements

The requirements for achieving discrimination and the degree to which it is required will vary with the nature of the system in question. In some systems discrimination predominates, e.g. certain transmission systems require specially matched and calibrated relays to achieve a close ratio between the time/current settings. In other systems, and particularly in medium-voltage systems such as supply networks or industrial installations where H.R.C. fuses are normally used, such a high degree of sensitivity is seldom necessary. These systems normally branch out from distribution points or busbars into sub-circuits such that the distributor fuse can more often than not be several times the current rating of the sub-circuit fuse. The ratio occurring between the two is usually of the order of 2 : 1 or more, and such applications account for the largest proportion of locations in which H.R.C. fuses are employed.

The H.R.C. fuse can, in fact, offer a greater degree of discrimination than 2 : 1 under certain conditions, and is often required to do so; but this does not alter the fact that the best engineering practice is to arrange the layout of the system to afford as great a margin as possible. Where a greater degree is required, a knowledge of the circuit conditions and fuse characteristics are necessary in order to make a correct choice. Whatever degree may be required, it follows that any protective device must be capable of giving consistent protection for the whole of its life in service.

3.2.3 Factors affecting discrimination

Fuse performance. H.R.C. fuses do not give close discrimination at very high short-circuit currents because of their properties of current and energy limitation which are the more important factors, but will do so at the lower values of overcurrent. The performance of an H.R.C. fuse designed to have the property of non-deterioration can be predicted for any given circumstances, because it remains faithful to its declared characteristics. The data required for fuse discrimination can be made readily available (see Sections 3.3.9–3.3.11).

System planning and layout. Discrimination can depend as much upon the design of the system as upon the performance of the protective devices involved, and the best results can be achieved only if the protective devices are co-ordinated with the system as a whole in the first stage of planning. An attempt to plan a distribution system merely on the basis of load requirements, with protective devices as an afterthought, will usually result in unnecessary complication and defeat its own object.

Non-deterioration, All forms of protection, whether discriminative or otherwise, depend for their effectiveness upon the manner in which their pristine condition can be preserved in service. They must either possess the property of complete non-deterioration or must be retested, maintained and recalibrated at regular intervals throughout their service lives. Since fuses are static devices without moving parts, they are not expected by the user to require maintenance and are seldom disturbed over periods of years unless a fault occurs. Non-deterioration is therefore one of the essential properties on which discrimination depends.

3.3 Fuse characteristics affecting discrimination

3.3.1 Fuse operation

Positive discrimination between any two H.R.C. fuses in series is achieved when the larger or 'major' fuse remains unaffected by fault currents which cause the smaller or 'minor' fuse to operate (or blow). When an H.R.C. fuse operates, the element absorbs energy from the circuit and heats up until it melts, vaporizes and disperses. This action is then followed by a period of arcing which persists until the resistance across the fuse builds up to a sufficiently high value to reduce the current to zero. The former period is known as the pre-arcing period and the latter as the arcing period (see Fig. 3.1). The heat produced within the fuse-link during operation is the integral $\int i^2 r \, dt$, where i is the instantaneous current and r the instantaneous resistance during the operating time.

3.3.2 Effect of fuse resistance

The resistance of two fuse-links chosen to discriminate with one another will be different, and each will change as the elements heat up during the pre-arcing period. The rates of change of r (i.e. dr/dt) during this time will also be different, but the rate of change in the minor fuse will be greater than that in the major fuse, so tending to assist discrimination.

The resistances of the fuses in relation to the resistance of the rest of the fault circuit are not usually very large. Up to the point where the minor fuse begins to arc, the change in the resistance of either fuse will not therefore affect the current appreciably. This is obvious from the oscillogram in Fig. 3.1, which shows the voltage drop across the fuse is very small during the pre-arcing period. Fuse resistance is a factor in discrimination, but only to a small degree, and since changes in resistance tend to assist rather than complicate the desired effect, it is convenient to ignore them for all practical purposes and to consider only I and t in calculations.

3.3.3 Conception of $I^2 t$

The choice of major and minor fuses can then be made on the basis of $I^2 t$ ($\text{amp}^2 \, \text{sec}$), because the same current flows through both fuses for the same time and $I^2 t$ is the factor common to both. Values* of $I^2 t$ for either the pre-arcing or arcing period (or for the total operating period, which is the sum of both of these) can be obtained from test oscillograms or by other means.

The melting of a fuse element is the point of no return. Even if the circuit is interrupted elsewhere, once the element has liquefied it cannot be restored to its former state or even become a stable conductor. Thus, for the purpose of discrimination the element of the major fuse must not approach this point. In other words, the total $I^2 t$ (i.e. pre-arcing $I^2 t$ + arcing $I^2 t$) admitted during the operation of

---

*I is the r.m.s. current during fuse operation and is deduced from $\int i^2 \, dt$ over the pre-arcing, arcing or total operating time as appropriate.

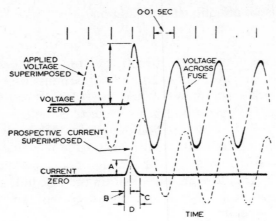

FIG. 3.1

Typical oscillogram of fuse operation on short-circuit, showing conventional terms as defined in B.S. 88: 1952

Fuse Performance		Test Circuit Conditions	
A. Cut-off current	29.2 kA	Prospective currents:	
B. Pre-arcing time	0.002 25 sec	r.m.s. symmetrical	47.6 kA
C. Arcing time	0.004 35 sec	r.m.s. asymmetrical	72.5 kA
D. Total operating time	0.006 6 sec	peak asymmetrical	102.5 kA
E. Maximum arc voltage	856 V	Applied voltage	440 V
		Power factor	0.13
		Frequency	50 Hz

the minor fuse must not exceed the pre-arcing $I^2 t$ of the major fuse. There must, in fact, be sufficient margin between the two that the major fuse element is not permanently affected by the $I^2 t$ which has to pass through it to blow the minor fuse. This principle is exemplified in Fig. 3.2, which shows $I^2 t$ values extended from test oscillograms of typical major and minor fuses.

3.3.4 The pre-arcing period

It is well known that H.R.C. fuses exhibit the property of cut-off, and at prospective currents at which cut-off occurs for a given fuse, the pre-arcing period is of the order of ¼ cycle. (Fig. 3.3 shows typical values taken from tests on an existing range of fuses.) For such short times it may be assumed that all the pre-arcing energy is absorbed by the element itself without loss to adjacent material and, being directly proportional to the mass of the element, is sensibly constant. The pre-arcing energies recorded empirically are lower than those calculated from the known physical constants of a particular element, owing to the effect of electromagnetic forces which cause the softened element to change form in the last stage of melting. At relatively low prospective currents, involving absence of cut-off and longer pre-arcing times, heat is lost from the element and the pre-arcing energy is consequently no longer constant but increases as a function of time. Fig. 3.4 shows the relationship between the prospective current and cut-off for a family of fuses referred to in

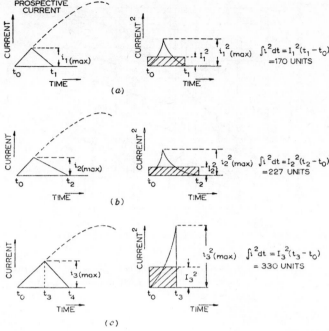

FIG. 3.2

Method by which $\int i^2 dt$ values may be extended from oscillograms of typical major and minor fuses: (a) minor fuse having an arcing time similar to its pre-arcing time; (b) minor fuse having an arcing time of twice its pre-arcing time; (c) major fuse.

$I_1 = $ (r.m.s. value of i_1)$t_0 - t_1$. $I_2 = $ (r.m.s. value of i_2)$t_0 - t_2$. $I_3 = $ (r.m.s. value of i_3)$t_0 - t_3$.

Since the total $\int i^2 dt$ of either minor fuse is less than the pre-arcing $\int i^2 dt$ of the major fuse, discrimination will be obtained

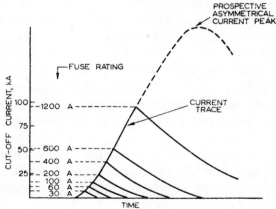

FIG. 3.3

Cut-off currents, for a given propective current, for a family of fuses

Prospective current:		Applied voltage	600 V
r.m.s. symmetrical	80 kA	Power factor	0.15
r.m.s. asymmetrical	100 kA	Frequency	50 Hz
peak asymmetrical	177 kA		

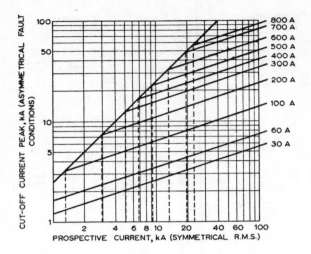

FIG. 3.4
Relationship between prospective and cut-off currents
for a family of fuses, and points at which
cut-off commences

Fig. 3.3 and indicates the prospective currents at which cut-off commences for each fuse.

3.3.5 The arcing period

The energy dissipated in the fuse during arcing consists of two components: one is derived from the inductive, or stored, energy of the circuit and the other is the energy fed into the arc directly from the source of supply.

The stored energy, $\frac{1}{2}Li^2$, is a function of the circuit inductance and the peak instantaneous current — which, in a fuse interrupting a heavy short-circuit, is the cut-off current. The inductance will differ for each circuit, as expressed in the power factor of an a.c. circuit or the time-constant of a d.c. circuit. The energy derived from the source of supply is a function of the instantaneous system voltage during the arcing period. For d.c. systems this is practically constant, but for a.c. systems it may vary from zero to peak and depends upon the point on the wave at which arcing commences.

The energy from both sources must be dissipated as heat within the arc. For given circuit conditions the design of the fuse can be varied to absorb the heat (and so cool the arc) at a desired rate. In the initial stages of arcing this can be achieved by varying the characteristics of the arc itself, and in the latter stages by varying the cooling medium. It must be assumed that the fuse has been designed and rated to keep the arc energy to a minimum consistent with the ability of the fuse to absorb the energy. In this connection it is important to ensure that the applied or system voltage does not exceed the rated voltage of the fuse.*

The circuit constants have as great an influence on the arcing energy as the characteristics of the fuse itself, and it will be appreciated that the variables introduced from both sources make the accurate prediction of arc energy a difficult problem. For a given fuse, however, it is possible to determine a maximum value of arcing I^2t which can occur. This is done in practice by calculating those circuit constants which will give the most severe arcing conditions and confirming from measurements taken from actual tests. For a.c. conditions the test involves applying the short-circuit at a point on the voltage wave which produces appreciable asymmetry of the current and at a critical value of prospective current (which is not necessarily the highest).

3.3.6 Short-time operation

When a minor fuse exhibits cut-off during the interruption of a heavy fault, the I^2t admitted during the arcing period is appreciable compared with that admitted during the pre-arcing period. In most cases the former exceeds the latter and is therefore a substantial part of the total operating I^2t.

It would be impracticable to take into account all the varying values of arcing I^2t when making a choice of minor and major fuses for the purpose of discrimination. The most convenient method and the safest is to assume the worst conditions and to consider only the maximum value, as explained in Section 3.3.5. If discrimination is possible at this maximum value, it follows that it will be obtained at the lower values which occur on the less-severe circuit conditions. It is important to remember that the maximum value would occur only under the most unusual conditions and would seldom be realized in service. The use of the maximum value therefore provides a useful margin of contingency.

3.3.7 Long-time operation

At the lower fault currents at which cut-off does not occur the arcing I^2t is small compared with the pre-arcing I^2t. As previously mentioned, the pre-arcing energy increases because of the increased thermal loss from the element, and at the same time the arcing energy is lower because the inductive energy in the circuit is less. Furthermore, under a.c. conditions the arc will tend to extinguish at voltage zeros and possibly at the next voltage zero after arcing commences. Thus, for operating times longer than approximately one cycle (0.02 sec at standard frequency) the arcing energy may be almost negligible, as illustrated in Fig. 3.5(a) and (b).

Since the arcing I^2t is small compared with the pre-arcing I^2t, it can be ignored for practical purposes and discrimination can be judged by merely comparing the

---

*If a well-designed fuse is used on system voltages which are within its voltage rating, the current decrement during arcing will be reasonably uniform. It can be shown that, where the pre-arcing and arcing currents follow uniform rates of rise and decrement, the arcing I^2t of a minor fuse may be twice that of the pre-arcing I^2t and still discriminate with a major fuse with double the rating of the minor fuse (see Fig. 3.2).

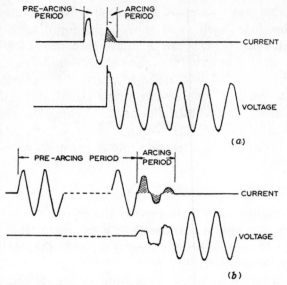

FIG. 3.5

*Examples of fuse operation showing proportions of
pre-arcing and arcing loops for longer operating times*

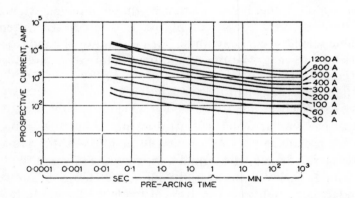

FIG. 3.6
Time/current characteristics

pre-arcing $I^2 t$ of the minor fuse with that of the major fuse. This is a simple matter of direct reference to the published time/current characteristics.

3.3.8 Two zones of operation

It is, then, apparent that the factors of H.R.C. fuse discrimination resolve themselves under two distinct headings:

(a) those involving low prospective currents which produce pre-arcing times longer than approximately 0.02 sec;

(b) those involving high prospective currents which produce pre-arcing times shorter than 0.02 sec.

The choice of fuses is made in terms of current rating. This is a convention which is necessary because of the first consideration in choosing fuses is to ensure that they are adequate to carry load under healthy circuit conditions. It does not always follow that the cut-off characteristics and $I^2 t$ values of a range of fuses conform to a regular relationship with respect to current rating. It is thus necessary to set down time/current and $I^2 t$ characteristics for each fuse in order to translate these data into terms of current rating.

3.3.9 Time/current curves

Fig. 3.6 shows typical time/current characteristics in which prospective current is plotted against pre-arcing time as prescribed in B.S. 88: 1952. For pre-arcing times longer than 0.02 sec it has been shown that the prospective current and the current admitted by the fuse are similar; thus, for a given current, discrimination depends only on the pre-arcing time and can be obtained between successive steps of current rating. For example, in a circuit where the prospective current is 3 kA the pre-arcing time of a 300 A fuse exceeds 0.02 sec and will discriminate with the next rating in the range, namely 350 A. The 350 A rating will discriminate with a 400 A one, and so on.

3.3.10 Values of $I^2 t$

Fig. 3.7 gives values of $I^2 t$ corresponding to each fuse rating and plotted in convenient form; from it the total $(I^2 t)_{max}$ for any minor fuse can be determined. Then, by extending this value to the curve representing pre-arcing $I^2 t$ values, the nominal rating of the smallest major fuse which will give positive discrimination can be read. The practical choice will then be the next larger standard rating. For example, the rating of the minor fuse is known to be 100 A in a circuit having a prospective current of 5 kA (which is above the value at which this fuse begins to cut off (see Fig. 3.4)). Fig. 3.7 shows the total $I^2 t$ of the 100 A fuse-link to be 105×10^3 amp^2 sec. The standard fuse-link whose pre-arcing $I^2 t$ exceeds this is a 200 A unit with a pre-arcing $I^2 t$ of 150×10^3 amp^2 sec.*

3.3.11 Virtual time

An alternative to the above method of determining discrimination is suggested in recent British Standards. This is based on the concept of 'virtual time', which is defined in B.S. 2692: 1956 as 'the time for which a steady current equal to the

---

*In the case of the fuses taken to illustrate the graphical method shown in Fig. 3.7 the pre-arcing values were first made to conform to a straight line by adjusting the scale of the ratings. It was then found that $(I^2 t)_{max}$ values taken from actual oscillograms also conformed approximately to a straight line. This indicates that for these particular fuses the relationship between pre-arcing $I^2 t$ and total operating $(I^2 t)_{max}$ follows a rational law.

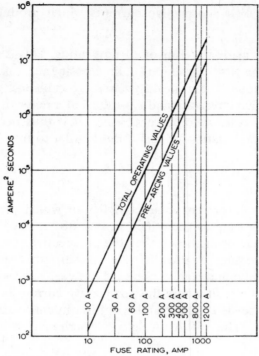

FIG. 3.7

$\int I^2 dt$ characteristics showing relationship between total
and pre-arcing $\int i^2 dt$ values for a family of fuses

prospective current would have to flow in a fuse to produce the same quantity of energy as would be produced if the actual current during the period of operation considered flowed in the fuse for the actual period'.

This time t_v is calculated from

$$t_v = \int i^2 dt / I_p^2$$

in which I_p is the prospective current and t_v is the virtual time of the period of operation under consideration.

Fig. 3.8 shows virtual-time/prospective-current curves for a typical range of fuse-links. These require an accurate measurement of $I^2 t$ from test oscillograms in the same manner as for Fig. 3.7.

To choose fuses for the purpose of discrimination, virtual-time/current curves are required for both the pre-arcing and total operating $I^2 t$ values of each fuse-link. The virtual total operating time of the minor fuse can then be compared directly with the virtual pre-arcing time of the major fuse after a reasonable tolerance margin has been allowed between the two. The theoretical merits of the virtual-time concept are somewhat overshadowed by the practical difficulty of presenting the curves in an easily usable form.

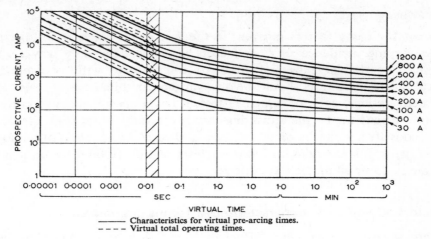

FIG. 3.8

Virtual-time/current characteristics. For times longer than 0.01 sec, the total operating curves tend to be pre-arcing curves and may be interpolated as shown in shaded region

3.3.12 Consistency of manufacture and quality

The theories so far expounded have been rationalized to some extent in order to meet practical conditions. The data on which the theories can be applied in practice are derived from type tests on actual fuses and on the assumption that the fuses used in service will be identical in characteristics with those used in the type tests. This in turn implies that the fuses used in service must be manufactured to a high standard of consistency. The accuracy of discrimination will therefore depend upon, and be proportional to, the consistency achieved. Consistency of characteristic is a function of quality control, but this in turn is dependent on design in as much as complication of design normally leads to difficulty in manufacture. Simplicity of design remains the keynote of consistency.

3.3.13 Manufacturing tolerances

Time/current characteristics are lines representing the mean value of tolerance bands, the width of which are a measure of the accuracy of manufacture. H.R.C. fuses can, with proper care, be made to conform to tolerance bands which are well within the requirements of normal service duty, but it is incumbent on the user to know the tolerance band when choosing fuses to give discrimination.

3.3.14 Non-deterioration during short-circuit operation

When a major and minor fuse are in series and the minor fuse blows, the major fuse must be able to handle the through short-circuit current without suffering damage. It will of necessity warm up, but provided that it cools down and resumes its pristine condition, it will then be able to discriminate in the future with a new replacement minor fuse. The stability of fuse characteristics under through short-circuit conditions is an essential factor and can be achieved by careful design.

3.3.15 *Proving tests*

In view of the many factors inherent in the fuse itself which affect discrimination and the varying incidence of such factors in service, the only way to prove the ability of a particular range of fuses to discriminate with one another is by empirical means. The tests should involve various combinations of major and minor fuse-current ratings, each of which is tested at various prospective currents from minimum current up to the rupturing-capacity rating. Other tests should take the form of consistency checks on stock fuses and careful inspection of unblown major fuses to check for non-deterioration. All tests require to be done on a sufficiently large scale to cover all possible variations.

3.4 Factors other than fuse characteristics affecting discrimination

There are several factors which affect discrimination other than those which are inherent in the characteristics of the H.R.C. fuses themselves, but they are not particularly critical and can be accounted for in system planning.

3.4.1 *Unequal loading*

Fig. 3.9 shows a hypothetical branching circuit in which the main fuse feeds a busbar from which are fed four sub-circuits. The aggregate of the current ratings of the minor fuses exceeds that of the major fuse by a considerable amount, but the aggregate of the loads on the minor fuses just about equals the full load on the major fuse. The circuit shown is exaggerated to illustrate the point and is unlikely to occur in this form in practice, but the tendency towards unequal loading is not uncommon and it is of interest to consider its effect. It can be surmised that the minor fuse, carrying the light load of 2 A (in Fig. 3.9), will be running almost at ambient temperature, whereas the major fuse, carrying full load, will be running warm. If,

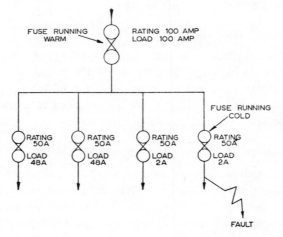

FIG. 3.9
Example of unequal loading which might affect discrimination

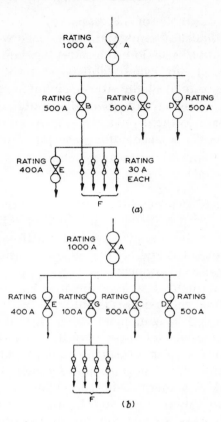

FIG. 3.10

Effect of system layout on discrimination. (a) B, C and D will discriminate with A, but E will not
   discriminate with B; (b) G is a new fuse, all fuses will now discriminate

then, a fault occurs on the load side of the lightly loaded minor fuse, a
discrimination ratio as decided from the fuse characteristics will be affected,
particularly for the longer operating times. Such cases do not occur very often in
practice and could usually be avoided by proper layout of the installation. Even
where they do occur there are several mitigating factors, because the difference in
running temperature between the two fuses is not as significant as may be imagined.
A proper evaluation of the difference can be made only by considering the running
temperature of the elements within the two fuses in relation to the melting
temperature of the metal from which those elements are made. For properly
designed fuses in which the elements are proportioned to run relatively cool, the
problem of unequal loading therefore becomes a minor factor.

3.4.2 *Effect of system layout*

Fig. 3.10(a) shows another circuit layout which occurs quite commonly in industrial
systems and is contrary to the best practice where a high degree of fuse

discrimination is either desirable or necessary. If it is assumed that for given circumstances in this particular system a ratio of 2 : 1 between the major and minor fuse is necessary to give discrimination, the sub-sub-circuit fuse of 400 A may not discriminate satisfactorily at the highest fault current with the distributor fuse at 500 A. The remedy here is to feed the 400 A circuit direct from the busbars, as shown in Fig. 3.10(b). The original sub-circuit distributor can the be reduced from 500 A to a size appropriate to the diversity on the remaining sub-sub-circuits. The economics of the change will vary according to site conditions, but should be easily assessable.

3.4.3 Prevention of abuse

The planning of all schemes of protection must proceed on the assumption that the equipment specified and installed is not unduly abused. When considering fuses this assumption may be safely made, because there is so little which can go wrong. Even fuses, however, must be installed and connected by electricians with sufficient skill to realize the importance of good connections.

H.R.C. fuse-links are normally tested in the fuse handles and fittings in which they are designed to be used in service. It is not uncommon, however, for fuse-links to be purchased and applied in equipment other than that for which they were originally designed. It is therefore a prerequisite of such applications that the terminals to which the fuse-links are attached should be adequate and similar in all essential respects to those which the fuse-link designer had in mind. It must be remembered that the link is a thermal device and must export some of the heat generated within itself to the contacts to which it is attached. Fuse fittings are normally designed on this basis, so that the link runs within the temperature limits required to ensure non-deterioration. It is prudent for the manufacturer to design fuse-links to carry their full load currents at temperatures well within the permitted limits, so as to allow an adequate margin for contingencies of service. If, however, there is blatant disregard of normal precautions, e.g. the use of extremely small terminals and undersized cables, it is only to be expected that the running temperatures will be other than normal and discrimination may suffer to some degree. It is hardly necessary to mention that the same trouble will be experienced if the connections are badly made, because these in themselves will be a source of heat which will cumulatively worsen until the point is reached where the fuse may blow prematurely.

3.5 Degrees of discrimination

It has been shown that the ultimate degree of discrimination which can be expected from fuses can be accurately defined by considering the characteristics of the fuses themselves and arranging the layout of the system with due regard for these characteristics. It has to be accepted that a rule-of-thumb ratio of the order 2 : 1 in current rating is required between the minor and major fuse for a relatively high degree of discrimination in systems where the fault level is high. For systems where the prospective currents are low and where the fuses do not exhibit cut-off, the ratio will not need to be so wide. On the other hand, for extremely important or vital circuits the ratio might need to be increased beyond 2 : 1. The point has been made that these ratios do not give any practical difficulty in system layout. Neither do they impose economic restrictions, provided that they are considered at an early stage in the system planning.

The importance of discrimination in relation to other factors in a particular system or installation may vary. Sometimes it is necessary to forgo some of the benefits of discrimination and to recognize the degree which is required or which can be obtained by a given choice of fuses.

3.5.1 Assessment of degree

The degree of discrimination required for any particular system is influenced by a number of factors, but these are not usually critical so far as the choice of fuses is concerned. It will therefore suffice to consider three possibilities:

(a) Those cases where the fuses protect vital circuits in which the continuity of supply and minimization of fault damage is essential.

(b) Less vital circuits where the consequences of loss of supply are not of first importance provided that the shut down is not prolonged.

(c) Cases where fuses are required mainly as back-up protection in independent circuits and where discrimination may not be the first consideration, provided that the fault is cleared without damage to equipment or the source of supply.

3.5.2 Vital circuits

Among the cases coming within the first category are such vital circuits as power-station auxiliary supplies, power supplies to continuous-process plants and circuits controlling vital services. In such cases the economy in capital expenditure is secondary to the requirements of safety or maintenance of the supply or both, and there can be no compromise in the choice of fuses or the layout of the system. The wise thing to do is to allow as wide a margin between major and minor fuses as the system will permit, and then to ensure that this is not less than that which the system requires. In this way discrimination can be assured with a sufficient margin of safety to account for all contingencies including the human element.

3.5.3 Practical solutions

Cases within the second category are becoming less easy to justify as industry becomes more dependent on electricity as a vital service. They usually occur through the rapid extension of existing systems which have not been planned to cater for such exigencies and where the cost of correcting the initial mistake makes it expedient to forgo the ultimate benefits of discrimination. There are those cases which occur by default because they are brought into being by practical electricians

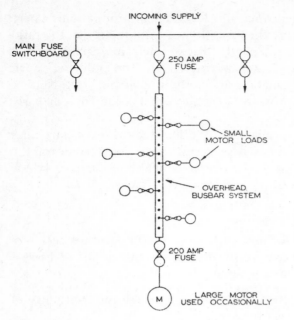

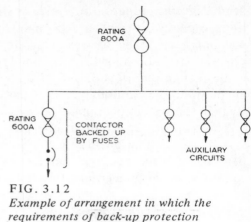

FIG. 3.12
*Example of arrangement in which the
requirements of back-up protection
predominate over discrimination*

FIG. 3.11
*Example of system in which a low degree of
discrimination may be expedient*

who are not familiar with the factors governing discriminative protection. The assessment of these cases is sometimes necessary after the event.

In such cases the degree of discrimination required will depend entirely upon economic factors. Installations in some industries are more fault-prone than others. Moreover, the speed with which the supply can be restored may vary according to the maintenance facilities available, and this will reflect on the losses sustained in production output or its equivalent. It must be appreciated that in all cases the lower degrees of discrimination can be tolerated if they result only in nuisance and do not involve danger.

In some instances discrimination may be foregone with proper discretion, as shown in the practical example given in Fig. 3.11. This illustrates an overhead busbar system which feeds a large number of small motors and is then looped off to a relatively large machine which is used only occasionally. There is no reason why the overhead busbar should not be used as the main distributor to such a motor; but if the fuses in the motor circuit are comparable with the main fuses, discrimination cannot be expected in the event of a heavy short-circuit in the region of the terminal of the large motor. Whether the consequences of this are important can be judged only in the individual case, but, since the consequences result only in inconvenience and do not jeopardize the safety of either personnel or property, the risk is easily assessed on technical considerations alone.

3.5.4 *Lighting circuits*

The lighting circuits in the same factory would be quite a different matter. Here, the consequences of general failure to discriminate might involve considerable danger to personnel if the factory were plunged into darkness while the machines were running. There is, however, no reason why the correct discrimination ratio should not easily be achieved for lighting circuits, because they consist mainly of small units of load evenly spaced and admirably suited for connection through branching sub-circuits of approximately equal size. Lighting circuits which do not discriminate are usually badly planned.

3.5.5 *Cases where discrimination is a secondary factor*

The third category covers only those specialized jobs where complete discrimination is known to be either inherently difficult or unimportant. Fig. 3.12 shows a typical instance. In this scheme the major fuse is required for clearing faults within the busbar zone. The size of the minor fuse is determined by the load of the particular sub-circuit which it controls.

Where such a scheme is adopted from expediency it is recognized that there is a band of fault currents at which discrimination cannot be expected. The minor 600 A fuse is necessary because it affords back-up protection within the through short-circuit capacity of the contactor with which it is associated, whereas the major 800 A fuse would not. Discrimination will, however, be obtained at the lower prospective fault currents, and because both fuses are relatively large, it will occur up to fault values of 15—20 kA (see Fig. 3.4).

3.6 Conclusions

H.R.C. fuses provide discriminative protection to a degree of sensitivity which is well within the requirements of normal distribution. Their characteristics can be ascertained and declared accurately, so that they can be applied in any circumstances where the requirements are known. A practical method of presenting the necessary data is by curves similar to Figs. 3.6 and 3.7, which give direct values and avoid the abstract complications of alternative methods.

H.R.C. fuses remain consistent and stable in service without calibration or maintenance, provided that they are properly designed and manufactured. The property of non-deterioration is one of the most significant factors in accuracy of discrimination.

The design of the system in which fuses are used and other factors external to the fuse itself affect the degree of discrimination which can be achieved. Assessment of these factors is necessary for the most effective use of fuses, but the assessment need not be critical.

Failure to achieve discrimination owing to incorrect choice of fuses may result in nuisance, but need not involve danger if both the major and minor devices are H.R.C. fuses of proper design and manufacture.

4 The role of the H.R.C. Fuse in the protection of low and medium voltage systems

The type of fuse which has come to be identified as the H.R.C. fuse was pioneered in the U.K. to meet the requirements of U.K. industry. The particular combination of high load density, system voltage and transformer capacity in this country had produced fault levels in low and medium voltage systems higher than those occurring in most other parts of the world. The emergence of the H.R.C. fuse together with the very large and sophisticated testing facilities to prove them put British fuse manufacturers in a leading position in both manufacturing and technology.

In the early 1960s détente with Soviet Russia improved to the extent that exchanges of technological information became possible and were officially encouraged. Fuse technology was considered to be an appropriate subject for such treatment and this paper was presented in Russia on the occasion of the British Trade Fair in Moscow in 1961.

Subsequently there has been a steady exchange of information leading to mutual benefits in trade and the establishment of mutual respect between individuals in this field of study.

Originally published in the *'English Electric' Journal*, September 1961.

4.1 Introduction

The high rupturing capacity (H.R.C.) fuse, as a method of obtaining reliable short-circuit protection at high values of fault current, was pioneered in Great Britain. An appreciation of the influences which preceded its introduction is helpful in the understanding of the present-day developments, which still continue at an undiminished rate.

Electrical engineers in the U.K. have been specially conscious of the need for efficient short-circuit protection because of the way in which electrical systems have evolved here. The U.K. became highly industrialized at an earlier stage than most other countries, so that the Electrical Industry grew rapidly from its earliest days. Such growth in a country of relatively small geographical area resulted in heavy and closely connected distribution networks and heavy load concentrations. In due course, co-ordination and interconnection was adopted on a national scale, involving the standardisation of transmission and distribution voltages. The system chosen for low voltage* distribution was 415/240 V, three-phase, a voltage somewhat higher than that used in most European countries or on the American continent. The high load densities also influenced the size of transformers feeding low voltage systems. These tended to be large and, in some industrial and city areas, were often of 1 and 2 MVA capacity in single or multiple units.

Under these circumstances fault levels grew to magnitudes which were not experienced in countries where electrical loads were distributed over larger geographical areas. Thus, the duty imposed on short-circuit protective gear increased at a rate which kept British engineers constantly aware of short-circuit problems on their low voltage systems.

The progress required in the field of electrical protection was intensified as electricity increased in importance as a service in industry and commerce, but other factors also conditioned the attitude of British engineers towards these problems. Long experience of industrial conditions had given rise to early legislation in the U.K. concerning safety requirements. These were accentuated by a strong public desire and demand for high standards of safety which, before the end of the nineteenth century had become embodied in the British Factory Acts. British codes of practice and industrial relations have been largely evolved round the Factory Acts and the electrical industry has from its earliest days taken the initiative in maintaining the highest standards of safety.

4.2 General trends of development

1920–30. In the early 1920s the position regarding fuse protection on supply systems became acute because of the rapidly increasing fault levels and because the then existing designs were proving no longer adequate for the duty imposed upon them. Up to that time many hundreds of ingenious ideas relating to fuse design had been patented in Great Britain. Most of these concerned the enclosure of or safe handling of wire or strip fuses mounted in various types of holders, including those of the simple cartridge type. The limitations of such fuses are well known and because of these limitations the H.R.C. fuse came into being as an urgent necessity in the more heavily developed areas of the country.

The term 'H.R.C. fuse', meaning high rupturing capacity fuse, was used from the first to distinguish the new conception in fuses from other designs with inferior performance.

The English Electric Company were among the pioneers in this field and quickly attained prominence with designs which not only satisfied the immediate requirements of the time, but also laid the foundation for subsequent developments which have proved their value up to the present time.

The first demand for H.R.C. fuses was from public supply authorities who required them for the protection of low-voltage networks. The nominal fault levels

*In the U.K. 415/240 V is officially classed as medium voltage, but the term low voltage is used in its general sense to include this voltage.

near to substations in highly developed areas was often in the region of 25 MVA at 415 V (three-phase equivalent). Faults on such systems caused wire or strip fuses and inferior cartridge fuses to fail explosively, causing uncontrolled flashover which resulted in more damage than that caused by the fault itself. In the worst cases equipment adjacent to that handling the fault was also wrecked and valuable services were needlessly interrupted. Fire hazards and danger to personnel also became serious. Thus, the first demand upon the H.R.C. fuse was for effective and safe rupturing capacity.

Rupturing capacity was achieved both by research and intensive empirical development. The results were then proved by exhaustive practical tests on the very systems for which the fuses had been designed. The supply authorities co-operated in this work with the fuse manufacturers by offering their systems as testing grounds and by allowing heavy short circuits to be deliberately applied for testing purposes. Such co-operation viewed in retrospect illustrates most vividly the gravity and urgency with which the problem was viewed at that time.

The particular requirement of the supply industry was that the new fuses should be made to fit existing fuse mountings in order to facilitate the rapid conversion of existing networks. This requirement severly limited dimensions and it is worth noting that the current ratings of fuses required for such conversions were of the order of 200 to 600 A. Fuse designers will appreciate that the difficulty of achieving rupturing capacity increases greatly with increase in current rating. It is to the credit of the earlier designers that they were able to produce the larger ratings with adequate rupturing capacity in such small dimensions even before the lower current ratings were considered. Fig. 4.1 shows a typical substation fuse distribution board which was converted in 1930.

1930–40. After some years of experience of H.R.C. fuses in the supply industry advantages other than rupturing capacity began to become evident. The main advantages were accuracy and fidelity of characteristics and the ability to resist deterioration even under adverse service conditions. These advantages commended themselves to the user industries which were becoming economically dependent upon adequate electrical protection as electricity rapidly supplanted other forms of power. It was for these reasons that during the 1930s the H.R.C. fuse became established as a major short-circuit protective device within industrial installations in the U.K.

The main factor which influenced this adoption was undoubtedly that of discriminative protection which was made possible by the accuracy of characteristics inherent in the new fuse. Experience quickly showed that co-ordination between H.R.C. fuses themselves and with other devices was predictable to a high degree. The economics implied in these facts obviously had a direct significance in terms of industrial efficiency.

Another factor which commended the H.R.C. fuse to the industrial user was its ability to remain stable and reliable over long periods of duty. The importance of

CENTIMETRES

FIG. 4.1

A substation board converted from open wire to high rupturing capacity fuses during the 1930s; (inset) The type of H.R.C. fuse developed for this purpose

this is apparent when the duty of the fuse is considered in relation to industrial usage.

An event of considerable importance in 1938 was the inauguration in the U.K. of the Association of Short-Circuit Testing Authorities. This is a voluntary co-operative organization set up by electrical manufacturers to provide a number of high power testing stations and formulate rules for the independent certification of circuit breakers, fuses and other short-circuit interrupting devices. The Association is supervised by government departments through the auspices of the National Physical Laboratory, by which means it is independent of individual manufacturers and serves the industry as a whole.

The emergence of testing facilities on such a large scale with adequate facilities for measurement and recording gave fuse designers an opportunity of proving their ideas and of pursuing further developments on a more scientific basis.

At this time H.R.C. fuses were beginning to fulfil a useful role in marine practice. Several passenger ships having large generating capacities were found to have higher

FIG. 4.2
One of several alternators installed at Nelson High Power Laboratory, Stafford

fault potentials than had been experienced previously. H.R.C. fuses were used to good effect and the opportunity this presented provided valuable experience on which present-day marine practice is based.

1940–50. In the 1940s the use of fuses and the scope of their application was greatly widened and accelerated due to the impetus of the Second World War. Quite apart from the natural growth in usage due to industrial expansion, varied applications arose in other fields.

Up to this point fuses and fusegear had been mainly used on industrial installations for protecting and controlling the distributor cables or sub-circuits. Busbar zones were in general protected by oil circuit-breakers. From 1940 onwards oil circuit-breakers began to give way at an increasing rate to combinations of fuses and manually operated airbreak switches (Fig. 4.5).

Another symptom of that period was the extent to which H.R.C. fuses were used to back up existing circuit-breakers to meet the increasing fault levels which had occurred since the circuit-breakers were first installed. This practice began as an expediency but quickly became recognized as an economic and logical arrangement for new equipment. Fuses required for both these applications tended to be larger in

FIG. 4.3
Power transformers installed at Nelson High Power Laboratory

current rating than had hitherto been necessary and ratings up to 1200 A became fairly common.

At the other end of the scale the smaller current ratings also increased in demand even on sub-circuits where the fault levels were not such as to require high rupturing capacity. This development was merely an extension of the benefits to be derived from accurate characteristics, discrimination and non-deterioration. Another advantage in using H.R.C. fuses throughout an installation was that calculation of fault levels in various situations was no longer necessary. Small fuses situated close to the main busbars could be of the same type as those at the end of the remotest sub-circuit. Thus, mistakes due to incorrect choice of fuses were eliminated.

During this decade the application of H.R.C. fuses in seagoing ships increased dramatically. The main reasons for this were still associated with rupturing capacity and the other factors mentioned, but an important new factor was the realization of the ability of the H.R.C. fuse to withstand considerable shock and vibration without sustaining damage to itself. The same factor also accounted for the increased popularity of fuses in the sphere of traction, particularly on railways and Underground tube systems.

Meanwhile, fault levels continued to rise in some spheres. It is a recognized tenet in electrical engineering that system fault levels should be kept as low as possible.

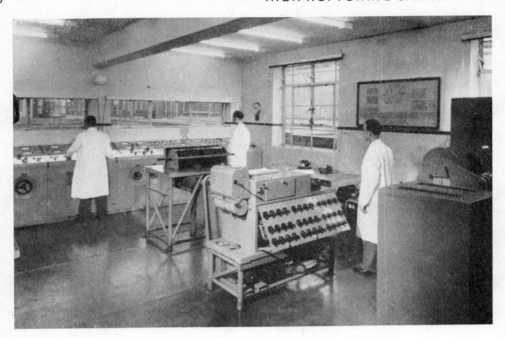

FIG. 4.4
A control room at Nelson High Power Laboratory

Cases began to arise however where system sectionalization became impracticable on the grounds of economy. For these cases H.R.C. fuses of still higher rupturing capacity were needed and duly produced. These practices were reflected in the current British Standards which prescribed categories of duty up to 35 MVA at 440 V for standard ranges of fuse-links. Special cases arose where 60 to 80 MVA capacities were needed and these also were met.

1950–60 Industrial recovery following the Second World War brought vastly increased usage of electricity in the early 1950s. The role of all protective gear intensified in importance. The effect of this on H.R.C. fuse development was to exploit its many advantages on a still more scientific basis. The ability of the H.R.C. fuse to limit short-circuit current and thus to minimize the effect of short-circuit stresses had long been recognized, but the emphasis up to that time had still been mainly upon rupturing capacity. The new circumstances meant that whereas in previous years engineers had been content if a protective device interrupted a faulty circuit without damaging itself, in 1950 they expected that it should also have the ability to minimize fault damage. The importance of this point of view was emphasized in many quarters and H.R.C. fuse designers were quick to exploit the new requirements. A considerable amount of work was done to explore and demonstrate the way in which short-circuit energy could be limited to minimize fire

FIG. 4.5
*A fuse switchboard incorporating fuses for protection
of busbar zone*

risk, danger to personnel, or the time required to repair a damaged piece of equipment in order to restore the service.

Other associated problems were those concerned with operational safety to personnel. Closer consideration was given to the possibilities of fusegear equipment being closed on to faulty circuits by unskilled operators. It was demonstrated that danger in such circumstances can be reduced to negligible proportions and the effectiveness of H.R.C. fuses when properly applied in this respect has now been amply proved by experience.

Fault energy limitation is also significant in relation to industrial automation and to the tendency towards large integrated continuous process plants. The effect of electrical faults on expensive capital plants can be economically serious. Any measures taken to improve the effectiveness of protection to prevent prolonged stoppages is obviously worthwhile.

One of the most exciting challenges to the fuse designer in recent times has been that of protecting semiconductor power rectifiers. The problem is difficult because of the extremely low thermal capacity of the recitifier cells. It has been stated that the ultimate output of these devices may well be determined by the effectiveness of the protective devices available. High speed H.R.C. fuses are now in regular and successful service for this duty, which has presented new problems entirely unlike any which have previously been encountered in fuse technology.

A further example of the increasing scope of H.R.C. fuses is that of the advances which have been made in the protection of aircraft systems. As in all other fields of electrical activity, generating capacities and therefore fault levels are increasing in aircraft at a rapid rate. System voltages in aircraft are also tending to increase and are not now dissimilar to those met with in industrial practice. Up to the present time it has been the custom to use conventional H.R.C. fuses in those aircraft circuits which required them. In 1960 the English Electric Company, after consultation with aircraft designers to properly ascertain their requirements, introduced a full range of H.R.C. fuse-links specially designed for aircraft use.

The art of H.R.C. fuse design has now progressed to such a stage that rupturing capacity in itself at low and medium voltages is no longer a serious problem. Rupturing capacities up to 200 kA r.m.s. (symmetrical) at 600 V are specified in some quarters and fuses are now available for these values. In this respect, the U.K. is in a much better position than most other countries because of the progress which has been made in testing station development, under the aegis of the Association of Short-Circuit Authorities, and by the enterprise of certain companies.

4.3 Fundamental parameters of performance

The basic technology of fuse design and performance is well known and has been discussed in published literature elsewhere. It is useful however, to reiterate the principles and to discuss the more fundamental aspects of performance, in order to better understand the prevailing points of view as regards H.R.C. fuse protection in the U.K.

4.3.1 Rupturing capacity

Rupturing capacity in an H.R.C. fuse stems from the fact that the fuse limits the short-circuit energy to a value within its own capabilities to absorb. The basis of fuse design is to predetermine the energy which will be released and to provide an arrangement of cartridge and filler which will contain this energy. Such information is usually interpolated from empirical data. The physical size of an H.R.C. fuse is roughly proportional to the nominal current rating and is largely independent of rupturing capacity. Hence, the smallest fuses can, if properly designed, have very high rupturing capacities.

The amount of energy required to melt a fuse is sensibly constant for very short operating or blowing times. Thus, however high the available or prospective short-circuit current may be the fuse will limit the actual current to a definite value, depending upon the rate of rise of the current. The rate of rise is maximum at the highest prospective current and maximum asymmetry. Hence, a limiting value can be defined which is associated with the severest conditions under which the fuse is intended to be rated. In British terminology this current is known as the cut-off current.

Electromagnetic stress within the fuse itself and also within the equipment it protects is proportional to $i_c{}^2$ (where i_c is the cut-off current).

Whenever fault current flows in a circuit inductive energy is stored and must be dissipated in the fuse or in the fault arc. (The circuit I^2r loss is usually negligible.) In the case of a solid fault all the energy is released in the fuse. Thus when the fuse elements melts arcing begins and releases the inductive energy. At the same time the system recovery voltage also attempts to sustain the arc. The dissipation of inductive energy in a fuse gives rise to a voltage $[L(di/dt)]$ across the arc. This is a function of fuse design inasmuch as the rate at which the arc is created and quenched is controllable. In the later stages of arcing the success of the fuse depends upon the rate of rise of dielectric strength being greater than the rate of rise of recovery voltage.

The pre-arcing or melting energy is proportional to $\int i^2 dt$ up to the time the fuse begins to arc. Likewise, the arcing energy is proportional to $\int i^2 dt$ during the arcing time. These terms are rationalised for convenience to I^2t where I is the r.m.s. value of i over the operating time of the fuse. I^2t also represents the thermal stress within the fuse and the circuit protected.

Rupturing capacity depends on keeping the total I^2t to a value which the fuse can physically withstand under all conditions of fault. For a given fuse element or arrangement of elements, the physical strength of the containing tube and its end caps must be sufficient to contain both the dynamic and thermal stresses created. The elements must also be so proportioned and disposed within the cartridge as to run cool on normal loads, and to be able to withstand transient overcurrents without damage. Designing to achieve this involves the control of the mass of the element which is consumed during arcing. The number and disposition of the elements is also a factor.

It is desirable to keep the pre-arcing I^2t to a low value but not necessarily to a minimum, as this must be balanced against the ability of the fuse to carry loads without overheating. It also must be chosen to provide proper discrimination with other fuses.

Arcing I^2t depends first upon the value of pre-arcing I^2t and then the means provided for quenching the arc. The arc is controlled by the shape and dimensions of the elements and also by the cooling media. Again the optimum value of the arcing I^2t must not be a minimum because of the considerations of arc voltage. Arc voltage must not be so high as to overstress the system insulation, bearing in mind that system insulation levels tend to reduce in service with the passage of time. The fuse operation should provide an even decrement of the current during the arcing period, so as to give an arc voltage well within the system insulation flash test value.

Another most important aspect of rupturing capacity which is often overlooked is that it determines the optimum accuracy required in manufacture. Thus, the accuracy required to give rupturing capacity ensures consistency and fidelity in other characteristics. This is one justification for using H.R.C. fuses in cases where fault levels are not particularly onerous.

4.3.2 Non-deterioration

Non-deterioration is a function of both fundamental design and control in manufacture. All protective equipment performs a passive role during most of its life and assumes an active role only when a fault occurs. This, on a well-found system, happens rarely. It is therefore necessary to know how a protective device may change its characteristics with the passage of time. With this knowledge it is possible to decide how often it must be changed or recalibrated to remain completely safe and effective. These questions have a special significance in the case of H.R.C. fuses because engineers do not expect to have to change them, even after many years of service. This is a natural attitude which arises because the fuse is a purely static device. Its elements cannot be inspected and it is practically incapable of being recalibrated once it has left the manufacturer's factory. The ideal fuses must therefore be designed in such a way as to prevent deterioration and so maintain safe characteristics for a long period. Account must also be taken of the possibility that the fuse may be subject to transient overcurrents of various kinds and to other hazards, even including incorrect usage. The property of non-deterioration must be regarded as one of the essential factors in the effectiveness of fuse protection.

Fuses taken from service after twenty-five years of continuous duty have proved that non-deterioration is an accomplished fact. Such duty has included continuous and cyclic loading, short-time overloading, high transient overcurrents, wide variations of atmospheric and ambient conditions, industrial contaminations and the hazards of handling and routine maintenance.

The features in design which determine non-deterioration are, broadly speaking, the correct choice of materials for chemical compatibility; correct heat transfer by proper proportioning of components; avoidance of dry joints and connections; proper protection of all exposed parts to withstand atmospheric pollution. To give effect to such designs involves scrupulous cleanliness and quality control in manufacture.

Non-deterioration also depends upon recognition of an optimum current rating and proper ratio between this and minimum fusing current. It presupposes that the fuse will not be made to carry currents greatly in excess of its nominal rating for prolonged periods in service. This does not necessarily mean that the ratio (or fusing factor) between current rating and minimum fusing current need be unduly high.

The art of fuse design therefore centres around the correct choice of parameters to give current rating, rupturing capacity and non-deterioration in adequate proportions.

As a corollary to non-deterioration the design should be such as to ensure that the fuse operates safely at all values of overcurrent below minimum fusing current. Such operation may take place after the fuse has been overloaded for very long times, perhaps after many weeks.

4.3.3 Discrimination

Discrimination under heavy fault conditions occurs between H.R.C. fuses when the total I^2t admitted through the smaller of two fuses connnected in series is less than

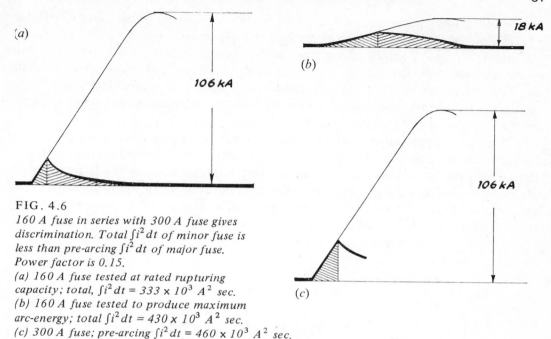

FIG. 4.6
*160 A fuse in series with 300 A fuse gives
discrimination. Total $\int i^2 dt$ of minor fuse is
less than pre-arcing $\int i^2 dt$ of major fuse.
Power factor is 0.15.
(a) 160 A fuse tested at rated rupturing
capacity; total, $\int i^2 dt = 333 \times 10^3 \; A^2 \; sec.$
(b) 160 A fuse tested to produce maximum
arc-energy; total $\int i^2 dt = 430 \times 10^3 \; A^2 \; sec.$
(c) 300 A fuse; pre-arcing $\int i^2 dt = 460 \times 10^3 \; A^2 \; sec.$*

the melting of pre-arcing $I^2 t$ of the larger fuse. As fuses are usually identified by
their nominal current ratings, it is necessary to know the appropriate $I^2 t$ values of
each. In the interest of simplicity it is normal to consider only the value of total $I^2 t$
which results under the severest arc energy conditions. If discrimination can be
obtained on the basis of such values, then it will be obtained under all conditions
which can occur in service. Fig. 4.6 shows oscillograms from actual discrimination
tests and illustrates the principles involved.

The factors inherent in fuse design which affect discrimination are: the ratio of
pre-arcing to arcing $I^2 t$; the magnitude of the fault current in relation to the cut-off
or limitation of current; and the accuracy of the fuse under all conditions. Other
factors external to the fuse itself, such as the layout of the system, can also
influence discrimination and must be taken fully into account, although the
calculations in respect of these need not be critical.

H.R.C. fuses will give discrimination to a degree which is well within the
requirements of normal service. Their characteristics can be predetermined and
declared with an accuracy which compares favourably with induction relays and
other similar devices.

4.4 British standards

British Standards exist for all types of fuses. Low- and medium-voltage H.R.C. fuses
are governed by B.S.88 which is the principal specification on which other standards

and specifications have been modelled. This standard concentrates on performance and categories of duty rather than dimensions. These are recommended but are not mandatory, a policy which has proved to be wise inasmuch as it has allowed full scope for development without the restrictions which are unavoidable if dimensions are made mandatory at too early a stage. Design has in fact crystallized due to the lead given by certain manufacturers. Thus standardization has occurred naturally by mutual consent of the users.

B.S.88 lays down current and voltage ratings; categories of short-circuit duty; fusing factors; limits of temperature rise; methods of presenting time/current characteristics; and conditions of usage. It also prescribes tests for the determination of all these on the basis of the most onerous conditions likely to be encountered in service. British Standards are constantly reviewed and amended as necessary to keep them up-to-date. B.S.88 is no exception and is currently being revised.

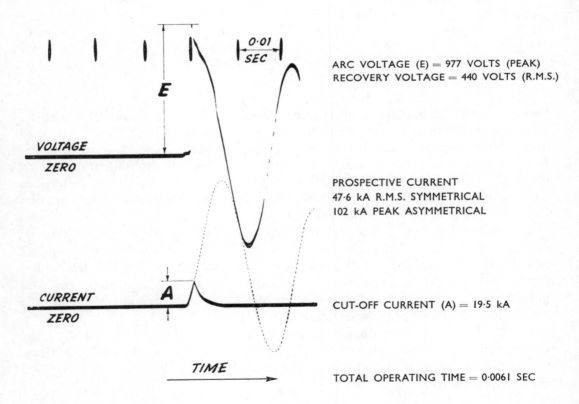

FIG. 4.7

An oscillogram of a test on an 'English Electric' fuse of 200 A 440 V rating: system, 35 MVA at 440 V three-phase; power factor, 0.13; closing angle, 31°; pre-arcing I²t, 160 x 10³ A² sec; arcing I²t, 500 x 10³ A² sec; resistance of fuse after operation, 500 MΩ

4.4.1 Proving tests and certification

It is usual for a manufacturer of H.R.C. fuses to present to the user proof that the fuses sold comply in all respects with the British Standard. In addition, he issues such data as the user may need to enable him to supply the fuse to maximum effect.

Proof of compliance of the fuse with the requisite standard is usually in the form of test certificates, which are records of type tests carried out under prescribed conditions. A typical example of a test certificate proving short-circuit performance is the 'Certificate of Short-Circuit Rating' issued by the Association of Short-Circuit Testing Authorities. This Association will test fuses only if they are designed in accordance with the relevant British Standard and comply fully with the criteria of proof laid down. The short-circuit test is performed at an authorized testing station under the supervision of an observer appointed by a government-sponsored laboratory.

Fig. 4.7 shows a typical test oscillogram which indicates the method of test and the conventional technical terms in fuse technology. It will be seen that the prospective current, power factor, closing angle (point on voltage wave at which circuit is closed), applied voltage, limits of arc voltage, and insulation of blown fuse are all prescribed.

The maximum arc energy within a fuse during short-circuit operation does not necessarily occur at the maximum prospective current for which the fuse is rated. B.S.88 provides for tests which produce the maximum arc energy condition in addition to tests at the higher prospective current.

Other test reports are issued by the fuse manufacturer in respect of temperature rise at full load, minimum fusing current, fusing factor, insulation resistance and time/current characteristics.

4.4.2 Fuse data

Fig. 4.8(a) shows time/current characteristics for a typical range of H.R.C. fuses. The prospective current is plotted against the pre-arcing time. The curves do not extend for times shorter than 0.01 seconds because at times below this the fuse begins to exhibit cut-off and a different relationship applies.

A direct way of presenting data in respect of the high prospective currents is to declare the $I^2 t$ values. The pre-arcing $I^2 t$ of a particular fuse is reasonably constant. The arcing $I^2 t$ varies with the circuit constants, but if the value is taken from the tests which produce the maximum arc-energy this is a limiting value which will provide for safety under all circumstances.

Fig. 4.8(b) shows $I^2 t$ values for a range of fuses, plotted in a convenient form for assessing discrimination as well as giving total values for use in relation to circuit protection. Figs. 4.8(c) and (d) show other data normally issued in respect of fuse application.

4.5 Applications of H.R.C. fuses

The applications of H.R.C. fuses are numerous and involve the employment of the

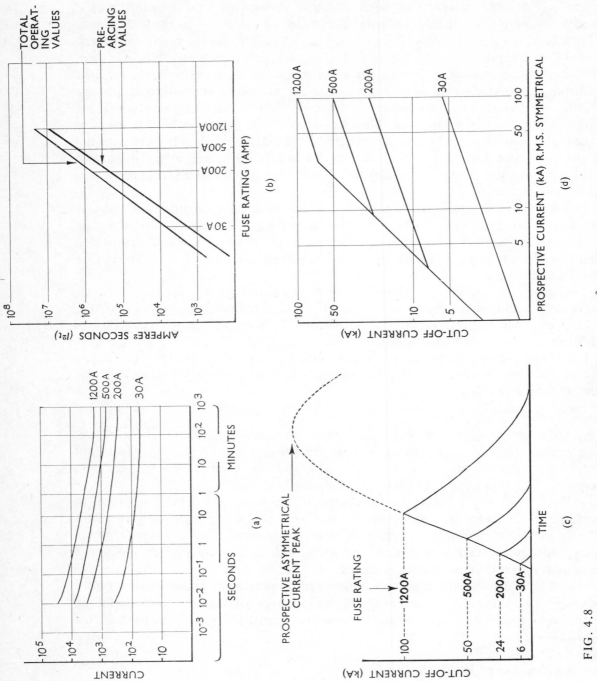

FIG. 4.8
Presentation of typical fuse data. (a) time/current characteristics; (b) I²t characteristics; (c) mode of operation; (d) current limitation

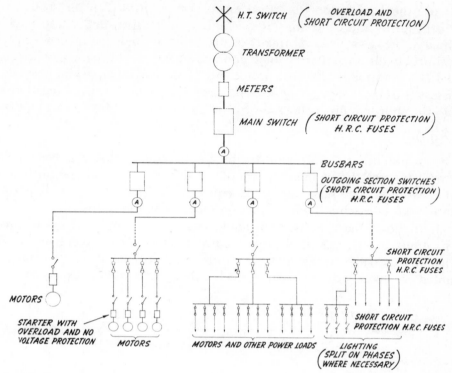

FIG. 4.9

Relative positions of short-circuit and overload protection in a typical industrial installation

principles of fuse design in varying degrees. For general industrial use the parameters are rationalized to give the widest scope of application. For special duties the parameters are defined to close limits.

The following examples are chosen to give a general picture of the whole field of application, but it would be impossible in the space available to give more than an outline sketch of the subject.

4.5.1 Protection of industrial distribution systems

The majority of H.R.C. fuses are used in industrial installations. In these a distinction is made between short-circuit faults which are electrical in nature and overloads which may arise from other exigencies. Short-circuit protection is required primarily at all positions in a system. Overload protection is primarily required at the point of consumption to protect the consuming device, as indicated in Fig. 4.9.

H.R.C. fuses in the first place are for short-circuit protection but embody a sufficient measure of overload protection to suffice for busbar zones, distributor cables and other conductors. Overload protection additional to the fuse is required to protect motors and other machines. Control of the distributors can be achieved by manual airbreak switches, normally of the 'on-load' type and incorporating fuses.

The short-circuit capacity of cables, contactor gear, and busbars may be related to the total $I^2 t$ admitted by the fuse protecting them. This relationship normally needs to be assessed from type tests because although the $I^2 t$ values of the fuses are readily available the short-circuit withstand of other equipment is not so easily obtainable.

4.5.2 Cables

It has been found that the $I^2 t$ admitted by a well-designed fuse is so small compared to the withstand of the equipment it protects that it is only in marginal cases that accurate data is necessary for such equipment. This is easily illustrated by considering the short-circuit withstand of cables in relation to fuses.

Fig. 4.10 shows the thermal withstand of cables in terms of $I^2 t$ compared with the $I^2 t$ admitted by fuses of the same current rating. The protective margin is so wide that it is possible to say that the short-circuit withstand of the cables may be disregarded completely on a system protected by H.R.C. fuses of this type.

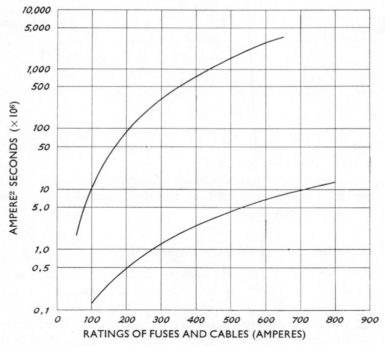

FIG. 4.10

Short-circuit protection of cables: upper curve, cable withstand;

4.5.3 Busbars

The same sort of relationship also applies to busbar structures except that in this case other factors emerge. Take for example a busbar of 1000 A rating from which are fed airbreak switches of 100 A rating. The interconnection between the busbar and the switch will normally be of the same rating as the switch but will be protected by the device which protects the busbar. If this device is a 1000 A H.R.C. fuse it can be easily demonstrated that the maximum $I^2 t$ which will be admitted by the fuse when a fault occurs in the switch will be well within the withstand of the 100 A conductors. These statements apply both to the electromagnetic stresses and to the thermal stresses. Fig. 4.11 shows the results of a test in which a 1200 A fuse

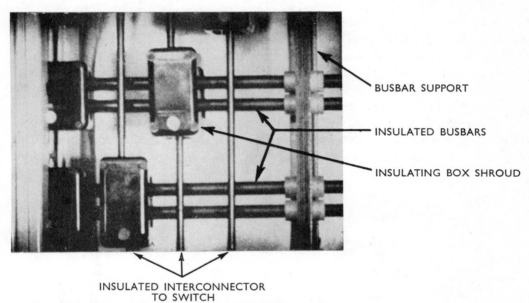

BUSBAR SUPPORT

INSULATED BUSBARS

INSULATING BOX SHROUD

INSULATED INTERCONNECTOR
TO SWITCH

FIG. 4.11
Photographs from a high-speed ciné film (1000 f.p.s.), showing conditions before (a) and after (b) a test on a 60 A interconnector protected by a 1200 A fuse

was blown when in series with such a conductor of 60 A rating. The values were exaggerated to show visible effects, but it will be noted that although the conductor is distorted it is still capable of continuing in service at least until such time as it is convenient to replace it.

Figs. 4.12 and 4.13 show designs of fuse switchboards in which these principles are incorporated.

4.5.4 Contactor gear

The back-up protection of motor control contactors introduces other factors. The vulnerable items are the contacts which, unlike static conductors, are liable to movement under stress and are subject to much higher current densities at the point of contact. Furthermore, they are liable to be 'closed' on to fault as well as to withstand 'through' fault currents. The fuses chosen for back-up protection must be co-ordinated in such a way as to allow the contactor to operate at all values of overcurrent within its own capacity. They must also be able to carry the transient over-currents during motor starting without deterioration. Thus, the current rating of the fuse may be twice or three times the rating of the contactor.

Most British manufacturers of good-quality control gear have carried out full-scale tests at 25 to 35 MVA on their combinations of H.R.C. fuses and contactors. The results show that the H.R.C. fuse limits $I^2 t$ to values well within the withstand of all

FIG. 4.12
Combination fuse switchboard for heavy industrial duty

FIG. 4.13
'Superform' fuse switchboard for industrial duty in cleaner situations than those envisaged for the switchboard shown in Fig. 4.12

reputable designs of contactor. Short-circuit protection is no longer a problem in this field, as is clearly demonstrated in Fig. 4.14.

4.5.5 Earth faults

Industrial systems in the U.K. are almost invariably three-phase 415/240 V with earthed neutral. The largest majority rely on H.R.C. fuses for system protection, including earth fault protection. This raises particular problems because the fuse ratings are in general larger than the cable ratings and have an average fusing factor

FIG. 4.14
Contacts of 100 hp contactor after the tabulated test (see Table 4.1)

TABLE 4.1

Summary of three-phase, short-circuit tests on motor contactors rated for motors of 7.5 hp, 25 hp, 50 hp, and 100 hp at 600 V

Unit	Duty	46 kA			66 kA			80 kA		
		Average prospective current (kA)	Maximum peak (kA)	Average recovery voltage (V)	Average prospective current (kA)	Maximum peak (kA)	Average recovery voltage (V)	Average prospective current (kA)	Maximum peak (kA)	Average recovery voltage (V)
7.5 hp	TC	47.9	3.88	567	68.5	4.63	571	79.8	5.6	570
	M	46.7	3.38	576	66.4	4.63	585	80.3	5.71	583
25 hp	TC	47.8	9.62	570	68.5	10.3	570	79.7	10.2	564
	M	46.5	9.7	570	66.4	11.26	585	80.7	11.2	583
50 hp	TC	47.9	17.3	566	69.1	18.7	572	80.4	19.8	568
	M	47.0	16.2	575	66.6	18.5	585	81.2	19.5	588
100 hp	TC	47.9	22.8	571	68.3	25.8	566	80.6	27.1	572
	M	47.6	22.8	577	66.3	25.5	585	81.8	28.6	594

TC, Through fault with contactor closed;
M, Making or closing contactor on to fault.

(i.e. ratio between current rating and minimum fusing current) of 1.6. Under these circumstances earth fault protection depends entirely upon the effectiveness of the earth return path to the star point of the transformer. That is to say, that the earth path impedance must be low enough to allow sufficient current to flow to blow the fuse quickly in the event of an earth fault. There is no difficulty in satisfying these conditions in the majority of industrial installations. The H.R.C. fuse provides a reliable method of earth fault protection which has all the advantages of simplicity and does not require specialized maintenance.

4.5.6 Low-magnitude faults

Consideration of earth faults brings into focus the possibility of faults of low magnitude. The limitation of fault energy by an H.R.C. fuse is applicable only on those faults which are of sufficient magnitude to cause the fuse to 'cut off'. At lower currents the time taken for the fuse to blow is in inverse proportion to the current. It is necessary to consider whether in the event of a low magnitude fault the fuse will blow before the system sustains damage.

As a general rule satisfactory protection can be obtained as exemplified in Fig. 4.15. In practice the possibility of a low-magnitude fault (as distinct from an overload) occurring in this context is slight because electrical faults in cables usually develop quickly into high-magnitude faults unless the cable impedance itself is too high. The fuse then deals rapidly with the developed fault and gives complete

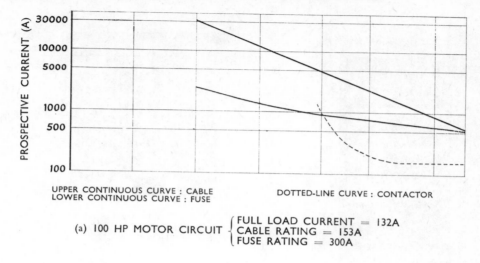

UPPER CONTINUOUS CURVE : CABLE
LOWER CONTINUOUS CURVE : FUSE

DOTTED-LINE CURVE : CONTACTOR

(a) 100 HP MOTOR CIRCUIT $\left\{\begin{array}{l}\text{FULL LOAD CURRENT} = 132A \\ \text{CABLE RATING} = 153A \\ \text{FUSE RATING} = 300A\end{array}\right.$

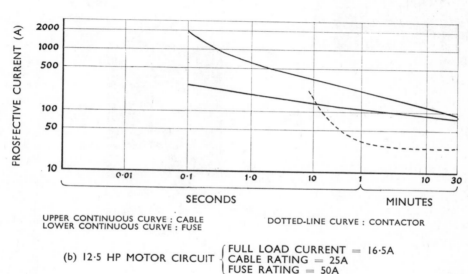

UPPER CONTINUOUS CURVE : CABLE
LOWER CONTINUOUS CURVE : FUSE

DOTTED-LINE CURVE : CONTACTOR

(b) 12.5 HP MOTOR CIRCUIT $\left\{\begin{array}{l}\text{FULL LOAD CURRENT} = 16\cdot5A \\ \text{CABLE RATING} = 25A \\ \text{FUSE RATING} = 50A\end{array}\right.$

FIG 4.15

Thermal protection of cables under moderate fault conditions. The cable curves indicate the 'no-damage' limit. The fuses afford protection, independently of the contactor, for overcurrents similar to those due to stalled rotor conditions; (a) 100 hp motor circuit: full load current, 132 A; cable rating, 153 A; fuse rating, 300 A (b) 12.5 hp motor circuit: full load current, 16.5 A; cable rating, 25 A; fuse rating, 50 A

protection. If the cable impedance is too high, this is an unusual condition and must be specially catered for.

Fig. 4.16 shows a typical design of distribution fuseboard incorporating fuses of ratings which are commonly involved in such problems.

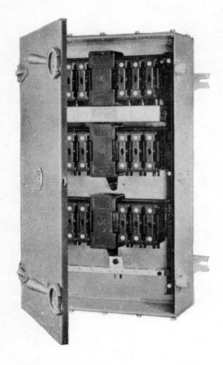

SHROUD REMOVED
FOR CABLING

BASE SECTIONED
TO SHOW SAFETY
SHROUDING

FIG. 4.16
An 'English Electric' Red Spot H.R.C. distribution board. The close-up views of fuse carriers and bases illustrate the operational safety features

4.5.7 Semiconductor rectifier protection

The outstanding characteristic of semiconductor rectifier diodes is their large output and relatively small size. This makes them inherently difficult to protect because their low mass involves low thermal capacity and low thermal inertia. They are therefore susceptible to damage from overcurrents even for very short times.

The only device that will respond to fault currents quickly enough to afford protection is the H.R.C. fuse, but this must be a specially developed type to match the characteristics of the semiconductor cells. A further complication is that these

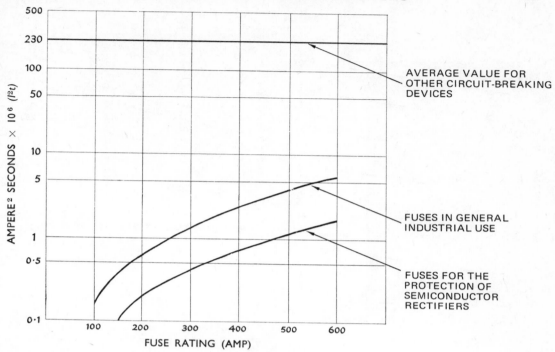

FIG. 4.17

Fault energy limitation of fuses in relation to other forms of circuit interruption

have a limited withstand against overvoltages. To be suitable for protection, a fuse must not generate arc voltage during fault operation because this would break down the reverse voltage withstand of the cell.

Fig. 4.17 shows the $I^2 t$ values, under given fault conditions, of a typical range of rectifier fuses compared to those of a typical range of industrial fuses. These curves demonstrate the order to which $I^2 t$ values can be reduced by specialized design, but there are several qualifications which must be made in achieving and applying them.

The worst condition that arises in practice is that which occurs when a cell in a recitifier bridge circuit breaks down·and allows the healthy cells of an adjacent phase to conduct through it. A fuse in series with the faulty cell must restrict the $I^2 t$ to a value within the withstand of the healthy cells. Therefore this withstand must be known. Fuses are then chosen on the basis of $I^2 t$ and this is an essential parameter in the rating of the fuse.

The chosen fuse will have an assigned normal current rating based upon its temperature rise limitations. Such ratings are determined by standard tests in which the fuses are connected to conductors rated at a given current density and mounted in free air. Correction factors may be applied for different mountings or ambient conditions. The rating of the chosen fuse may well determine the full load rating of the cell and its associated equipment.

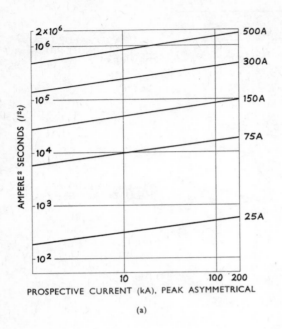

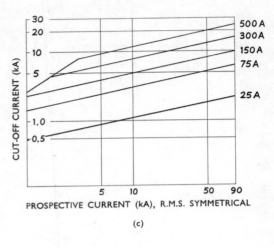

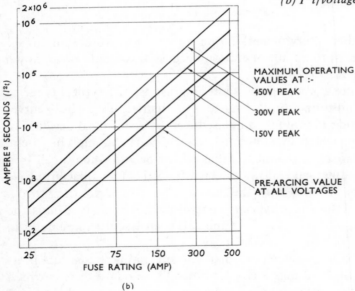

FIG. 4.18

*Typical characteristics of fuses for the protection of
semiconductor rectifiers:*
(a) I^2t/prospective current characteristic;
(b) I^2t/voltage characteristic; (c) current limitation

SEMI-CONDUCTOR
RECTIFIER PROTECTION

INDUSTRIAL

VOLTAGE RATINGS: 150, 300 AND
(PEAK INVERSE)   450 VOLTS

VOLTAGE RATING: 440 VOLTS.
(R.M.S.)

RUPTURING CAPACITY :
UP TO 200kA PEAK ASYMMETRICAL

RUPTURING CAPACITY :
UP TO 250kA PEAK ASYMMETRICAL

UP TO 150 AMP.   32 GRAMMES.

UP TO 100 AMP.   205 GRAMMES.

UP TO 350 AMP.   152 GRAMMES.

UP TO 300 AMP.   470 GRAMMES.

UP TO 700 AMP.   307 GRAMMES.

UP TO 500 AMP.   1200 GRAMMES.

TYPE GS   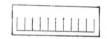   TYPE T

CENTIMETRES

FIG. 4.19

A comparison of size and weight of fuses used for semi-conductor rectifier protection and those for industrial use

The fuse must then be considered in relation to the applied voltage required from the equipment. Again, a semiconductor cell is seldom used in service at voltages closely approaching its ultimate voltage withstand. Fuses are designed to produce arc voltages not exceeding twice their own voltage ratings without damage to the cells being protected. The arc voltage across a fuse is $L(di/dt)$, and being a function of the length of the arc is virtually independent of the applied voltage. This means that a fuse designed for a high applied voltage affords no protection to a cell rated for a much lower voltage. Hence it is necessary to have a choice from a range of fuses of various voltage ratings as well as at various current ratings.

The next factor to be considered is that of rupturing capacity. Semiconductor rectifier equipments tend to be large in some industrial applications and the prospective fault currents are correspondingly large. Fuses must therefore be

FIG. 4.20
100 A silicon cells individually protected by 300 A fuses in 750 kW
equipment

designed for high prospective current values and proved by test. 250 kA (peak asymmetrical) is not an unusual value.

A variety of rectifier arrangements exist and fuses may be used in different positions in such arrangements. The duty of the fuse under both load and fault conditions will vary according to its position in the circuit.

Finally I^2t will vary as a function of both the applied voltage and prospective current. Where these differ from the tested values the true service values must be obtained by interpolation. Curves are necessary for this purpose and Figs. 4.18(a), (b) and (c) are typical of the kind of data which is now available. Fig. 4.19 shows typical fuses compared to industrial fuses. Fig. 4.20 shows fuses fitted into a rectifier equipment.

4.5.8 Aircraft fuses

The increasing fault levels on some modern aircraft and the adoption of higher voltages have made the introduction of better short-circuit protection imperative. The H.R.C. fuse has been adapted for this purpose and offers many advantages not available by alternative means.

The parameters of performance required in aircraft are unlike any which occur in other applications and it has been necessary to design fuses specifically for aircraft use. Obviously minimum weight is a ruling factor but this must be related to other

AIRCRAFT

VOLTAGE RATING : 250 VOLTS.

RUPTURING CAPACITY :
UP TO 42kA PEAK ASYMMETRICAL

UP TO 20 AMP.   5 GRAMMES.

UP TO 50 AMP.   13·5 GRAMMES.

UP TO 100 AMP.   26·5 GRAMMES.

UP TO 160 AMP.   64·5 GRAMMES.

UP TO 3C0 AMP.   126 GRAMMES.

TYPE UA

MARINE

(ADAPTED FOR AIRCRAFT)

VOLTAGE RATING : 440 VOLTS.

RUPTURING CAPACITY :
UP TO 84kA PEAK ASYMMETRICAL

UP TO 15 AMP.   5 GRAMMES.

UP TO 30 AMP.   14 GRAMMES.

UP TO 60 AMP.   32 GRAMMES.

UP TO 200 AMP.   146 GRAMMES.

TYPE AP

INDUSTRIAL

VOLTAGE RATING : 440 VOLTS.

RUPTURING CAPACITY :
UP TO 250kA PEAK ASYMMETRICAL

UP TO 30 AMP.   58 GRAMMES.

UP TO 60 AMP.   77 GRAMMES.

UP TO 100 AMP.   205 GRAMMES.

UP TO 200 AMP.   280 GRAMMES.

UP TO 300 AMP.   470 GRAMMES.

TYPE T

CENTIMETRES

FIG. 4.21

A comparison of size and weight of fuses for aircraft, marine (adapted for aircraft) and industrial use

FIG. 4.22
H.R.C. fuses installed in a contemporary aircraft

aspects of performance. The rupturing capacity specified by the aircraft industry at the present time is 16 500 A r.m.s. symmetrical at 250 V a.c. and 16 500 A at 120 V d.c.

In the U.K., aircraft fuses must comply with specified characteristics and withstand tests laid down by the Air Registration Board. These include: vibration tests over wide frequency bands, climatic, mould growth, tropical exposure, acceleration, crash landing and endurance tests. Also, in order to rate the fuse for various positions, either outside or inside the pressurized zones of the aircraft, research has been necessary regarding the effects of high altitude. The results of these developments can be seen in Fig. 4.21.

Close attention to the characteristics required in relation to motor-starting and other peculiarities of aircraft electrical practice has shown that the H.R.C. fuse is equal to other forms of protection in this respect. Moreover, the property of fault energy limitation becomes exceedingly significant in the case of aircraft protection. In civil aircraft, in particular, the minimization of fire risk from cables subjected to short-circuit stresses is vital to the extent that no smoke or smell must be evident to alarm passengers. Research is still proceeding in this direction but the results so far achieved are in themselves sufficiently clear to commend the H.R.C. fuse for widespread applications. Fig. 4.22 shows an actual installation in a modern aircraft in which the specially-designed fuses are in use.

5 Protecting armoured P.V.C. cables

The introduction of P.V.C. as an insulating material for low-voltage industrial cables virtually revolutionized the design of industrial distribution systems. The changes affected the basis of rating for cables because they brought a significant reduction in thermal withstand under overload and fault conditions. Regulations were amended to cater for the protection of the new cables and these tended to throw some doubt on the ability of many existing H.R.C. fuses to meet the changed circumstances. Conversely it was made to appear that the use of existing fuses might have resulted in the uneconomic use of the cables.

These assumptions were wrong, partly because the existing specifications for fuses did not adequately emphasize their capabilities to protect against overload damage and partly because the relationship between fuse protection on the one hand and cable withstand on the other had been assessed on academic rather than practical criteria. Urgent investigations and tests were carried out to show that existing H.R.C. fuses were quite adequate to protect the new P.V.C. insulated cables. They established the reasons for the apparent misunderstandings and brought standards and regulations into line.

This work was subsequently reflected in I.E.E. wiring regulations and British Standards for fuses and provided impetus for much closer and practical liaison between fuse and cable designers, a liaison which has proved very valuable on many subsequent occasions.

Originally published in *Electrical Review,* October 1962.

5.1 Introduction

In the U.K. the majority of distributor cables are protected by H.R.C. fuses of the type which come within class 'Q' of B.S. 88:1952. Class 'Q' fuses are defined in the British Standard as those having fusing factors between 1.25 and 1.75, but a large proportion of them are designed towards the middle value and have fusing factors of about 1.5. In general terms, therefore, the indications have been that armoured multi-core P.V.C. insulated cables will withstand currents of about 50 per cent above their ratings for a sufficient length of time to permit a H.R.C. fuse to blow.

This experience has been somewhat at variance with published information concerning the ratings of P.V.C. cables, but recent researches have shown that the conclusions drawn from practical experience are valid and that the cable ratings first put forward have been overcautious.

When it was realized that there was an apparent contradiction between service experience and the published rating tables for P.V.C. cables investigations were started to evaluate the position. These investigations have been carried out by various interested parties and this chapter deals with those made by the English Electric Co. Ltd.

5.2 Practical tests

To put the subject into perspective and to pave the way for more objective tests, a series of practical tests was first carried out under simulated service conditions. Various lengths of standard armoured multi-core P.V.C. insulated cable were connected to standard distribution fuseboards and protected by fuse-links corresponding to the current rating of the cables. The cables and fuses were then subjected to a variety of overload conditions from 125 per cent (of the nominal cable rating) upwards. In each case the current was maintained until the respective fuse-link operated. Many of the cables were subjected to several tests in succession, some of which included cyclic overloading.

Measurements of current and temperature on the fuses and cable cores at various distances from the fuses were continuously recorded throughout each of the tests. After each test the cables were examined for signs of damage. The cables were inspected externally and tested electrically and then they were laid open for visual examination. Samples of the insulation were tested for hardness and subjected to a variety of mechanical tests. These and other measures were taken to assess by practical judgment whether the cable would have been fit for continued and unrestricted use.

Samples from a wide range of sizes were tested. It was noticed that the majority of samples in all sizes did not attain the limiting temperature of 120°C. The few which did were nevertheless unaffected by the extra temperature above 120°C. Although these tests could not be accepted as conclusive, there was no doubt that they showed that a very large proportion of applications in service would be free from hazard. The tests were satisfactory from every practical point of view. No cables were damaged even where the rating of the fuse-link exceeded that of the cable.

Thus, the cables seemed to be much better as regards temperature withstand than had been officially recognized. Furthermore, the fuses behaved better than a literal interpretation of 'fusing factor' would suggest, in that the heat from the cable or other sources interacting on the fuse-link had the effect of adjusting fusing factor towards closer protection.

It is now known from other sources that a new overload limit of 50 per cent has been proposed instead of the 25 per cent previously adopted. This means that class 'Q' fuses having fusing factors not exceeding 1.5 will provide adequate protection without qualification. The new limits do not refer to new developments in P.V.C. but to the reassessment of its behaviour in existing designs of cable. Thus, the new

FIG. 5.1
*200 A fuse mounted in free air and arranged in accordance
with B.S. 88, but with extra cable lagging to produce a
conductor temperature of 70°C at rated current*

overload limits now recognized apply equally well to cables already installed, provided they comply with B.S. 3346:1961.

5.3 Fuse behaviour

For safe protection a margin is required between the survival limit of the cable and the fuse characteristics. The margin inherent in a fusing factor of 1.5 is due to the fact that fuse characteristics tend to speed up when the fuse is used in service under temperature conditions which are different from and higher than those obtaining when nominal fusing factor is determined by test. This is a natural phenomenon which is common to any equipment which is responsive to thermal influence. It is accepted that cable ratings may vary according to whether they are installed in air or in the ground, or run singly or bunched. Fuse ratings also vary, but since the fuse may be installed in circumstances which are different from those which govern the cable rating, the amount of the variation will also be different.

A further series of tests were therefore carried out to study the effect upon fuse characteristics due to variations in environmental conditions, and to determine whether the margins thus obtained would ensure the protection required. The tests were designed to produce an analysis of the factors which affect the ability of a H.R.C. fuse to provide overload protection.

Fig 5.1 shows a typical test arrangement in which a 200 A fuse fitting containing a 200 A fuse-link is mounted on an open panel in air to ascertain its time/current characteristics under these conditions. The test complies with B.S. 88, but the cable has been lagged to produce a conductor temperature of 70°C on full load to

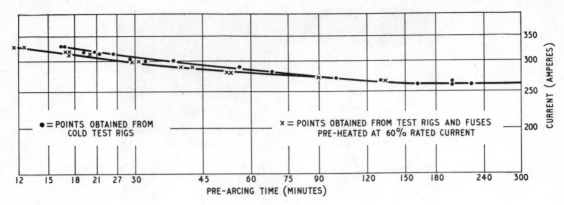

FIG. 5.2
Time/current curve for 200 A fuse-link in a 200 A fitting in free air, when connected to 0.1 in^2 cable lagged to produce a conductor temperature of 70°C at rated current

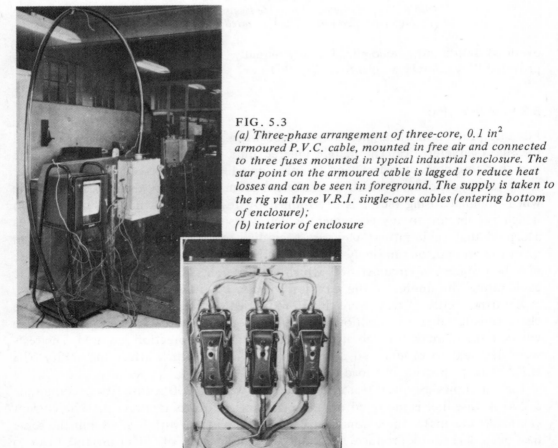

FIG. 5.3
(a) Three-phase arrangement of three-core, 0.1 in^2 armoured P.V.C. cable, mounted in free air and connected to three fuses mounted in typical industrial enclosure. The star point on the armoured cable is lagged to reduce heat losses and can be seen in foreground. The supply is taken to the rig via three V.R.I. single-core cables (entering bottom of enclosure);
(b) interior of enclosure

simulate the conditions envisaged for multi-core P.V.C. insulated cable. Tests were conducted under two conditions:

(a) With the fuse starting from cold, i.e. having carried no load and at ambient temperature before the overload currents were applied;

(b) Starting from warm, the fuses having been pre-heated for several hours by current of about 60 per cent rated current.

Fig. 5.2 shows that the effect of pre-heating has no significance for times of longer than about 75 min. Although there is a divergence for shorter times, this is not important because the fuse gives an ever-widening margin of protection at all currents which cause the fuse to blow in less than about 75 min (see Fig 5.6).

Tests were then carried out as shown in Fig. 5.3 (a) and (b) with three fuses in one enclosure and connected to a three-phase load by a 0.1 in² multi-core armoured P.V.C. cable. As for the previous tests, the fuses, and fuse-links were of 200 A rating. After a full range of time/current tests had been taken with this arrangement, it was

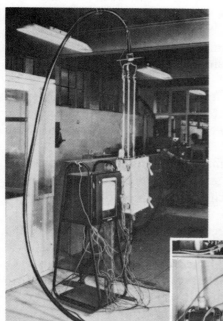

FIG. 5.4
(a) Similar to arrangement in Fig. 5.3, except that the P.V.C. armoured cable is connected to the fuse via ⁹/₁₆ in diameter copper risers, 3 ft long;
(b) interior of enclosure

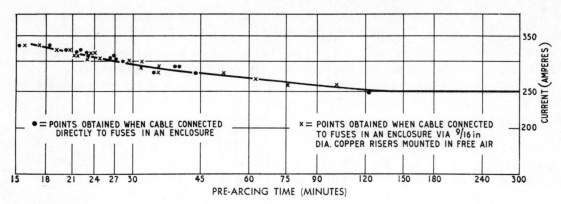

FIG. 5.5
Three-phase tests on 200 A fuse connected to 0.1 in², three-core, P.V.C. armoured cable

varied to the extent that solid copper rods were interposed between the fuses and the cable ends as shown in Fig. 5.4 (a) and (b). The rods had a cross section of double that of the cable and were about 3 ft long. The intention here was to simulate the conditions which can occur in circuits where the fuses are relatively remote from the cable and are connected by means of other conductors or through busbar arrangements.

Fig 5.5 shows that the difference between the two cabling arrangements is small enough to be ignored. It is calculated that the effect on time/current performance due to the variation in conductor size would, however, become significant when the cable tails inside the enclosure were of considerable length.

The question of whether the test arrangement for mounting the fuses is representative of all service conditions may be arguable, but it represented most conditions where fuses are used in fuse switches and is not dissimilar to the condition which applies in distribution boards.

Fig 5.6 shows the results of the tests in perspective with the relevant part of the published time/current curve, which in this case is representative of a 200 A fuse-link fitted into a 300 A fuse fitting. The published curve has been in existence for many years and the test conditions upon which it was based represent the more usual conditions which obtained in service at the time it was issued.

Also inset with this curve is a curve representing the temperature of a P.V.C. cable core corresponding to the various points on the fuse time/current characteristic. It can be seen that the temperature approaches the limiting value only under those overload conditions which would cause the fuse to blow in a relatively long time. Thus, overloads which are only slightly greater than the minimim fusing current of the fuse cause the latter to operate in a much shorter time with corresponding lower cable core temperatures.

The tests show that the most significant effect on fuse characteristics is due to the nature of the enclosure and these findings are consistent with those of an E.R.A.

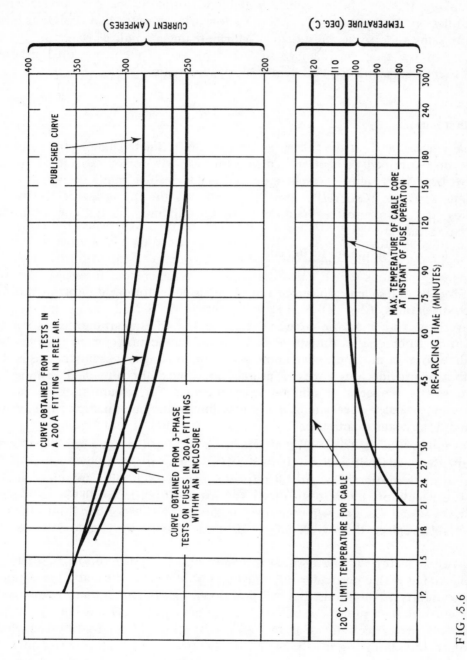

FIG. 5.6

The upper curves show the characteristics of a 200 A fuse-link under various conditions of service; the lower curve shows the temperatures attained by the cable in relation to time/current (fault energy) let through by fuse (mounted in free air)

report published some years ago. Other tests have shown that the relative size of the enclosure does not greatly affect the results. The effect of the fuse fitting is also noticeable but this is a factor which would not normally occur in service where P.V.C. cables are used. It would be pointless to work cables to their economic limits and at the same time to use fuse fittings which are uneconomically large. In no case during any of the tests was damage sustained by any of the cables used although many of them were subjected to several tests in succession.

5.4 Fusing factor

It is usually accepted that the fusing factor is an approximate guide to the ability of a fuse to give overload protection. This assumption has proved satisfactory in the past, but the closer limits which apply to P.V.C. cables make it necessary to understand what the term really implies. In B.S. 88 the fusing factor of a fuse is defined as 'the ratio, greater than unity, of the minimum fusing current to the current rating, namely:

$$\text{fusing factor} = \frac{\text{minimum fusing current}}{\text{current rating}}\text{'}$$

B.S. 88 does not specify optimum values of fusing factor but groups fuses within broad classifications. Class 'Q' H.R.C fuses, which are the subject of this chapter, may have any fusing factor between 1.25 and 1.75 and the onus is on the manufacturer to choose an optimum value between these limits which gives the best results in service. It must be borne in mind that the principal function of the H.R.C. fuse is that of providing short-circuit protection; it must also provide discrimination and possess the property of non-deterioration so as to remain stable in service without maintenance. Overload protection is important but must not in any way compromise the other functions.

The optimum fusing factor is that which allows the fuse to permit legitimate temporary overloads but causes the fuse to blow within the survival limits of the equipment to be protected when 'illegitimate' overloads persist. In some circumstances it can be more dangerous to have too low a fusing factor than one which is too high, because if fuses blow prematurely under normal service conditions the user will inevitably replace them with larger fuses and nullify the overload protection altogether.

Since fusing factor varies according to conditions of service, any value of fusing factor associated with a particular fuse must be regarded as a nominal value obtained under specified test conditions. B.S. 88 lays down that the minimum fusing current of fuses must be determined 'in one of the containing cases, if any, in which they may be used in service'. Incidentally, this stipulation relates to fuses which by definition include both the fuse-links and the fittings in which they are mounted. It is necessary to differentiate between 'fuse' and 'fuse-link' because the manner in which the fuse-link is mounted may affect its minimum fusing current.

Fuse-links are sold as separate entities and they may be used under a great variety of conditions. It is clearly not practicable for the fuse-link manufacturer to declare fusing factors for circumstances of which he has no precise knowledge. The nominal fusing factor which he decides upon must, however, be appropriate to actual service conditions.

In the past this has been done very successfully and the values adopted have in general been amply confirmed by service experience. For general distribution practice, a value of between 1.4 and 1.5 has been very widely used and is now shown to be compatible with the protection of P.V.C. cables.

5.5 British standard 88

Although the methods for determining fusing factor laid down in B.S. 88 have been satisfactory for practical purposes, it should not be overlooked that these allow considerable latitude. If, therefore, fuses are required to provide overload protection, the manner in which fusing factor is derived may need to be re-examined.

B.S. 88 states that 'For each fuse, two currents shall be determined, one of which is 90 per cent of the other, at the larger of which the fuse element shall melt within the appropriate time indicated in Table 5.3, and at the smaller of which the fuse element shall not melt within that time. The minimum fusing current shall be deemed to be the arithmetical mean of the two currents so determined'.

The following example illustrates the theoretical latitude which is open to a manufacturer within the meaning of B.S. 88. Take a given fuse which is designed to blow just within an 'appropriate time' of two hours at a current of 131 A but which takes slightly longer than two hours at 130 A. To determine minimum fusing current such a fuse could first be tested to ensure that it blows within two hours at 144 A,

$$\text{i.e. } \frac{130 \times 100}{90}$$

and, secondly, to show that it does not blow at 130 A within the same period. The mean of the two values is 137 A and this may be taken as being the minimum fusing current according to the method laid down in B.S. 88. However, the same fuse is known to be able to blow at 131 A within the stated period and not to blow at 117.9 A (which is 90 per cent of 131 A). The mean of these two values is 124.7 A, which is an alternative minimum fusing current still within the meaning of B.S. 88.

If the fuse is now tested to find out the maximum current which it will carry within the temperature rise limits laid down to ascertain the maximum permissible full-load rating, and if this value were to be found to be 100 A, then the fusing factor could be either 1.37 or 1.247 according to the preference of the manufacturer. But there is a further complication because the manufacturer is under no obligation to rate the fuse at 100 A. This is a maximum figure which he can assign but not the minimum one. He may choose advisedly to allow the fuse to run cooler

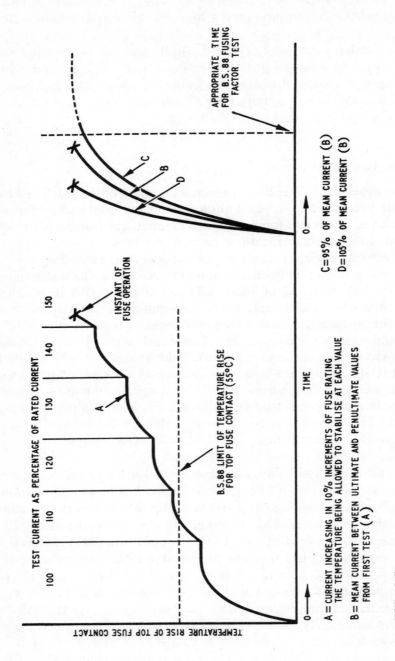

FIG. 5.7
Typical method for determination of minimum fusing current

TEST CURRENT AS PERCENTAGE OF RATED CURRENT

100   110   120   130   140   150

TEMPERATURE RISE OF TOP FUSE CONTACT

A

INSTANT OF
FUSE OPERATION

B.S.88 LIMIT OF TEMPERATURE RISE
FOR TOP FUSE CONTACT (55°C)

A = CURRENT INCREASING IN 10% INCREMENTS OF FUSE RATING
    THE TEMPERATURE BEING ALLOWED TO STABILISE AT EACH VALUE

B = MEAN CURRENT BETWEEN ULTIMATE AND PENULTIMATE VALUES
    FROM FIRST TEST (A)

C = 95% OF MEAN CURRENT (B)
D = 105% OF MEAN CURRENT (B)

APPROPRIATE TIME
FOR B.S. 88 FUSING
FACTOR TEST

TIME

C
B
D

at rated current by assigning a lower figure, say 80 A. On this basis and having regard to the method of obtaining a value for minimum fusing current, the fusing factor may then be any value between 1.56 and 1.71. Thus, for exactly the same fuse B.S. 88 permits the choice of any fusing factor from 1.247 to 1.71.

This example, which admittedly is an extreme one, merely shows that fusing factor, as it is at present defined, may not be a suitable basis upon which to specify the ability of a fuse to provide overload protection, unless the methods by which it has been derived are known. Reputable manufacturers do not, of course, invoke these extremes and are willing to declare the basis upon which the minimum fusing current of any particular range of fuses is obtained.

5.6 How fusing factor is obtained

The method of obtaining fusing factor which has been used by the English Electric Co. Ltd for many years complies with both the letter and spirit of B.S. 88 and at the same time produces values of minimum fusing current and fusing factor which can be directly related to service conditions.

A fuse-link is first designed by calculation for a desired current rating. It is then set up in the fuse fitting in which it is commonly used in service and mounted on a vertical panel. Cables of the appropriate size and length are connected to it and arrangements made to record temperatures at the fuse contact and elsewhere. Full-load current is then passed through the fuse for a sufficient time for the temperature to stabilize. The temperature rise (of the top fuse contact) thus obtained must not exceed the limit of 55°C laid down in B.S. 88. The test current is then increased by 10 per cent steps and at each step the temperature is allowed to stabilize. This process is continued until the fuse-link blows. Another similar fuse-link from the same batch is then put into the circuit and blown at a current which is a mean between the last two successive steps of current applied in the first test. If, for example, the temperature on the fuse contact levels out at 140 per cent current and the fuse finally blows at 150 per cent current, the value chosen for the second test would be 145 per cent. The time taken for the fuse-link to blow at this current is then recorded. Two further tests are then taken on similar samples, one at five per cent above the mean current and the other at five per cent below it. The times recorded in the latter tests are then used to ensure that the fuse-links comply with the 'appropriate time' laid down in B.S. 88:1952; the time corresponding to the higher current must be less than this time and that corresponding to the lower current greater.

The virtue of this method is that the value of minimum fusing current obtained represents an actual current at which the fuse will blow in an actual time and can thus be translated directly in terms of time/current characteristics. Fig. 5.7 illustrates a typical time/temperature curve from which the true mean value of minimum fusing current can be derived in the manner described. Fig. 5.7 also shows that the limiting temperature rise at full load is not exceeded. It thus gives an indication of the margin between the B.S. 88 limit and the actual fuse-link temperature, this being a measure of the ability of the fuse-link to cope with any additional external heating effects which may occur in service. Fusing factor derived in this way is, therefore, on an easily defined basis which can be directly related to service conditions.

5.7 P.V.C. cables and fuses in practice

The conclusions to be drawn from service experience and practical tests in the laboratory leave no doubt that H.R.C. fuses having nominal fusing factors not exceeding 1.5 will protect armoured P.V.C. insulated cables.

This presupposes that fusing factor has been determined on a realistic basis and the onus is upon the fuse manufacturer to satisfy the user that this has been done. If fusing factor is to be retained as a valid term for the purpose of assessing overload protection, it seems likely that the methods laid down in B.S. 88 will need to be reconsidered and possibly revised in a future issue. Another step which could be taken in this direction is the adoption of a standardized design of test rig on which all fuse-links could be tested to ensure that fusing factors declared by various manufacturers conform to the same principles.

When this has been done, it will still be necessary to recognize that the declared fusing factor cannot be other than a nominal value because it is liable to be influenced by service conditions. Fortunately, this is not serious because the fuse characteristics tend to adjust themselves in the right direction in situations where they are called upon to protect P.V.C. cables against overload. In these situations also fuse-links having nominal fusing factors of 1.5 provide a considerable margin of safety which allows for latitude in the choice of fuse ratings for a particular location. It is not always possible, for instance, to choose a fuse of a rating which comes within that of the cable. Fuses are rated to rationalized values, whereas cable ratings are variable according to the situation in which the cables are installed. The fact that the rating of the fuse-link may not coincide with that of the cable does not usually give rise to practical difficulty because the difference in ratings comes within the margin of safety.

Armoured multi-core P.V.C. cables have many attractions and it is an advantage to industry to be able to use them without irksome restrictions. It is also important that established practice relating to fuse protection and based on experience and comprehensive tests should be found to be compatible with the protection of the new cables. The distribution of electricity is a service to industry and not an end in itself. The simpler the techniques of distribution the more effective is the service likely to be. There are doubtless many points to be studied in the fields of both fuse and cable technologies, but the average user should not be expected to concern himself with these. He may take comfort in the fact that the techniques with which he is already familiar will achieve the results required.

6 Discrimination between H.R.C. Fuses in practice

This paper was produced as a natural corollary to the previous paper outlining general principles. Practical proof of any aspect of short-circuit performance can only be produced by testing at full power, that is to say, in testing stations capable of delivering the same order of short-circuit power that could occur in actual service under 'worst case conditions'. The testing also had to be done on a statistical scale sufficient to eliminate possible errors and this is a burdensome and costly undertaking.

Such tests were duly carried out and showed conclusively that the general principles can be obeyed and that the margins of safety previously recommended are sufficient to cover all contingencies.

Determination of a discriminating margin on the basis of I^2t let-through is a convenient rationalization. It allows for a considerable safety margin by simply ignoring the effect of resistance and the relative rate at which the resistance changes between two fuses intended to discriminate.

The point is made quite clearly in the records of the tests which also qualify the safe limits. Thus it is possible to apply closer coordination between fuses where this is necessary.

The final proof of discrimination is that the major fuse should not have suffered any change of state. A good deal of attention was paid in the investigations to this point and an interesting fact emerged. This was that fuses of the design in question will resist deterioration even when they are called upon to conduct values of I^2t closely approaching the known melting value. This was further evidence of the ability of fuses to carry 'through short-circuit' currents without suffering deterioration.

Originally published in the *'English Electric' Journal*, January 1963.

6.1 Introduction

The general principles governing discrimination between H.R.C. fuses are well established. The more important issue which concerns the user is how these principles apply to the fuses which are actually installed in service.

Over many years, investigations have been carried out to show how 'English Electric' H.R.C. fuses discriminate in practice. Information of this sort is continuously kept up to date and efforts are constantly being made to establish new criteria for discrimination to further improve distribution practice. Experience has shown that to be completely valid such criteria need to be obtained by testing under simulated conditions of service. It is impossible to calculate all the variables and fortuitous variations which occur in practice.

Fuses are normally chosen for discrimination from commercially published curves and other data which is itself obtained by empirical means. The validity of such data is as important as the fidelity of the actual fuses put into service. This is further justification for carrying out discrimination tests on practical lines. Moreover, such tests need to be comprehensive enough to ensure that fuses of every current rating in a range follow a consistent pattern of behaviour. They must also cater for the most severe circuit conditions which can occur and include short-circuit duties at full voltage up to the maximum for which the fuses are rated. It should be realized that the proof of performance for one range or type of fuse cannot be taken as establishing the pattern for all fuses, because contemporary designs vary over a wide field and differ considerably in characteristics.

Testing of this order requires test plant which is capable of producing not only the full value of prospective short-circuit power, but is also capable of controlling circuit conditions and providing accurately recorded measurements.

In practice, if two fuses appear to have discriminated with each other, the criterion of discrimination is not merely that one is open circuited and the other is not. It is essential for the unblown fuse to continue to give satisfactory service after the fault has been removed and the minor fuse replaced. The characteristics of the unblown fuse should therefore remain completely unimpaired. This implies the need for a safety margin beyond the theoretical limits which govern discrimination, to account for manufacturing tolerances and the fortuity of circuit conditions. As it is obviously not practicable to examine every fuse which has survived a through fault in service (this would involve a shut-down, which is the very thing discrimination is intended to prevent), this margin must be inherent in the method used for choosing fuses. Experience has shown that the relatively simple methods which have become established for choosing H.R.C. fuses in practice are satisfactory. Recent investigations show why this is so and also indicate the factors which determine the necessary safety margins and show how these are achieved in practice.

This chapter first reiterates the theoretical principles upon which discrimination depends and outlines some of the factors affecting practical applications. It then describes the methods by which the recent investigations have been conducted by drawing on examples from a series of tests carried out on 'English Electric' H.R.C. fuses, type 'T'. The tests described, which are representative of the whole series, prove that discrimination is obtained with a working margin such that the characteristics of the major fuse remain unchanged. It is of interest to note also how closely the performance of these fuses can be predicted from the published curves.

6.2 The principles of fuse discrimination

Discrimination may be defined as the ability of fuses to interrupt the supply to a faulty circuit without interfering with the source of supply to the remaining healthy circuits in the system. This requires that a larger fuse nearer to the source of supply will remain unaffected by fault currents which would cause a smaller fuse further from source of supply to operate.

When a fuse operates (or blows) it must first melt and then quench the arc which is created by the stored energy in the circuit and afterwards maintained by the system recovery voltage. The energy required to melt the fuse element is referred to as the pre-arcing energy, while that released in the fuse during arcing is known as the arcing energy.

Energies in this sense are dependent upon the square of the current (i^2), the time (t) and the resistance of the fuse. All these values are variables but as i and t are common to both of the fuses in series it is permissible to consider the effect upon the fuses in terms of i^2 and t. The energy absorbed in a fuse during fault interruption is proportional to

$$\int_{t^0}^{t^1} i^2 \, dt$$

which in common fuse parlance is rationalized to I^2t (i is the r.m.s. value of the current derived from the integral).

When an H.R.C. fuse blows it operates very rapidly and limits the current. Thus, it limits the quantity I^2t and this may therefore be regarded as the value which the fuse permits to pass or 'lets through'.

For discrimination the I^2t let through by the smaller fuse which blows must be less than the pre-arcing I^2t which would cause the large fuse (element) to melt.

Fuse-links are normally identified and chosen for duty by their current ratings, thus when choosing them for discrimination it is necessary to know the minimum ratio between the respective current ratings at which discrimination can be achieved.

This ratio may vary according to the magnitude of the fault current which is likely to occur. At higher values of fault current the fuse 'cuts off' or limits the current and interrupts the fault in less than half a cycle. The arcing I^2t under these conditions is a considerable part of the total I^2t let through and may be larger than the pre-arcing I^2t. It can be shown that for discrimination under these conditions the larger fuse-link must be about twice the rating of the smaller fuse-link, i.e. a ratio of 2:1. At smaller fault currents the arcing I^2t is small compared with the pre-arcing I^2t, and the ratio between the fuse current ratings may be much closer.

Discrimination is also influenced by several factors external to the fuse itself, but these are to a large extent self compensating and are often offset by the safety margins inherent in published data. This is particularly so for data relating to I^2t because this is based upon the most severe theoretical conditions, which are unlikely ever to actually occur in service. Thus the published data already allows for exigencies such as unequal loading and external thermal effects (e.g. heat from adjacent equipment) which may affect fuse performance. In most cases there is no difficulty in providing for an adequate ratio for discrimination if attention is paid to the planning of the system in which the fuses are to be used. Where, in unusual circumstances, the system layout does not permit a ratio of 2:1 between fuse ratings, the prospects of obtaining discrimination at any lower ratio must be assessed by reference to published data and due regard to the actual fault level of the point in the system where the fuses are situated. A lower ratio is often permissible, particularly when the fault level is low, or even at relatively higher fault levels where fuses of the larger current ratings are used e.g. a 500 A fuse will discriminate with a 400 A fuse at fault levels less than about 6 kA or 7 kA (5 MVA). This is because a ratio as high as 2:1 is only necessary for those fault levels above which a fuse begins to exhibit 'cut-off' and a typical 400 A fuse does not cut off below about 7 kA.

6.3 Discrimination tests

The work involved in investigating discrimination parameters can best be illustrated by considering the results of actual tests. The tests described here are chosen as being typical of those involved in a test series carried out at the Nelson High Power Laboratories to prove the discriminative capabilities of a range of 'English Electric' industrial H.R.C. fuse-links. The test procedure adopted was to take a number of current ratings as the major reference fuses and then to choose by means of published data the minor fuse-links with which the major fuse-links could be expected to discriminate under various fault conditions.

The range of fuse-links in question was tested for this purpose up to 35 MVA at 440 V, which corresponds to fault currents up to 46 kA (r.m.s. symmetrical). Other tests were carried out in the same series at 100 kA (r.m.s. symmetrical), also at 440 V.

The results obtained in the series of tests relating to the 100 A reference fuses are typical of those obtained throughout the range and serve to illustrate the factors involved. The reference fuses were tested against several ratings of minor fuse-links at differing fault levels and under a variety of other service conditions.

The minor fuse-links for the higher fault current tests were selected initially by reference to $I^2 t$ curves (illustrated in Fig. 6.1) which are issued in this form by the fuse manufacturer. For discrimination a minor fuse-link must be one having a total operating $I^2 t$ smaller than the pre-arcing $I^2 t$ of the 100 A fuse-link. It will be seen from Fig 6.1 that the 60 A fuse-link is a possible choice but, as it is near the limit, it was considered more practical to choose the next standard rating of lower value, namely 50 A.

Table 6.1 shows examples from the programme for the higher fault current tests. It will be noted that the minor fuse (a) is blown at a prospective current of 46 kA when connected in series with major fuse (A). Then follow three separate test shots at 46 kA in which the minor fuses (b, c and d) are blown successively against the same major fuse (B). Other tests are then recorded at a value of fault current which produces the maximum arc energy within the minor fuse, all the shots being recorded oscillographically.

An examination of the oscillograms shows quite clearly what processes are involved. Fig. 6.2 is a typical oscillogram of a 50 A fuse clearing a fault of 46 kA when in series with a 100 A fuse. The test conditions are those of B.S. 88:1952 which lays down the power factor and point-on-wave at which the fault must be

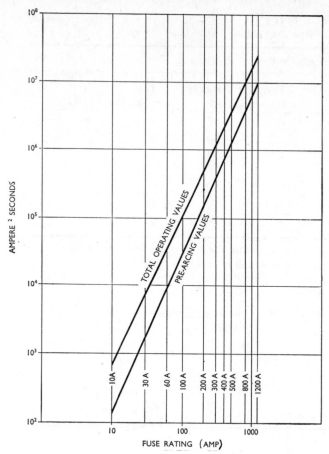

FIG. 6.1
$I^2 t$ characteristics, showing relationship between total
and pre-arcing $I^2 t$ for a range of fuses

initiated so as to produce the severest condition beyond those which are ever likely to be encountered in service at this prospective current.

The total $I^2 t$ ($\int i^2 dt$) of the 50 A fuse measured from the current trace of the oscillogram is 29×10^3 amp^2 sec, and the pre-arcing $I^2 t$ of the 100 A major fuse taken from the oscillograms of similar tests on this fuse is 34×10^3. Since the former value is smaller than the latter this satisfies the theoretical condition upon which discrimination first depends. However, practical success can be judged only after a thorough examination of the major fuse, to ensure that no deterioration has occurred which could impair its future performance. The results of these examinations are dealt with later.

It is a well-known phenomenon in fuse technology that the maximum energy released in a fuse when it blows does not necessarily occur at the highest prospective

TABLE 6.1

Discrimination tests at higher fault currents

Major fuse-link		Minor fuse-link		Prospective current (kA)	
Rating (A)	Test reference	Rating (A)	Test reference		
100	A	50	a	46,	100
		50	b	,,	,,
100	B*	50	c	,,	,,
		50	d	,,	,,
100	C	50	e	1.8†	
100	D	50	f	,,	
100	E	50	g	,,	

Same major fuse-link for three successive test shots.
†*Critical value which produces maximum arc energy.*

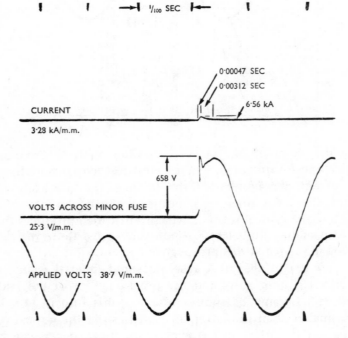

FIG. 6.2
*Discrimination test at 46 kA between 50 A minor fuse (a)
and 100 A major fuse (A) (see Table 6.1). Prospective
current: 46 kA (r.m.s. symmetrical), 106 kA(peak); voltage,
415 V; power factor, <0.15; frequency, 50 Hz; total I^2t of
minor fuse-link, 29×10^3; pre-arcing I^2t of major fuse-link,
34×10^3*

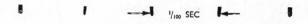

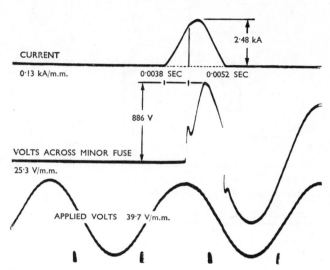

FIG. 6.3
*Discrimination test at 1.8 kA between 50 A minor fuse (c)
and 100 A major fuse (C) (see Table 6.1). Prospective
current, 1.8 kA (r.m.s. symmetrical), 3.86 kA (peak);
voltage, 415 V; power factor, <0.15; frequency, 50 Hz;
total $I^2 t$ of minor fuse-link, 32×10^3; pre-arcing $I^2 t$ of
major fuse-link, 34×10^3*

current (or highest value of MVA) corresponding with the maximum rupturing capacity at which the fuse is rated. The severest test may occur at some lesser value of prospective current, depending upon the design of the fuse. If discrimination tests are to represent the most onerous conditions which can occur in service it is necessary to test at the critical prospective current which produces the highest value of $I^2 t$. In this respect the critical prospective current refers to the minor fuse, since this is the one which actually operates.

Fig. 6.3 shows a typical oscillogram of the maximum arc-energy condition as it applies to the 100 A and 50 A fuses in series. The total $I^2 t$ of the 50 A fuse in this case is 32×10^3. This is still less than the pre-arcing value for the 100 A fuse and thus satisfies the theoretical condition required for discrimination. The incidence of the maximum arc energy condition in practice is fortuitous and depends on the nature of the fault. It is unlikely that it will be experienced more than once in a lifetime of any one major fuse and it may never occur at all. If therefore the 100 A fuse can be shown to have resisted deterioration during this test, there is no doubt that it will be able to resist the effects of any other faults (whether higher or lower in terms of current) occurring during the rest of its lifetime.

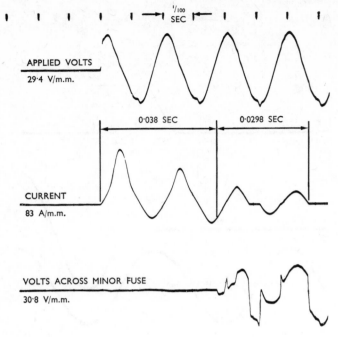

FIG. 6.4
*Discrimination tests at low fault current between 80 A minor fuse (h)
and 100 A major fuse (F) (see Table 6.2). Prospective current, 0.72 kA
(r.m.s. symmetrical), 1.78 kA (peak); voltage, 415 V; power factor,
<0.15; frequency, 50 Hz; total I^2t of minor fuse-link, 30 × 10³;
pre-arcing I^2t of major fuse-link, 34 × 10³; pre-arcing I^2t of minor
fuse-link, 27 × 10³; arcing I^2t of minor fuse-link, 3 × 10³*

The values of total I^2t shown in Fig. 6.1 are values derived from tests at critical prospective currents, i.e. those which produce the maximum arc energies in the fuses represented. This means that fuses chosen from Fig. 6.1 will discriminate successfully under the most severe short-circuit conditions that can theoretically arise. In practice it is unlikely that the worst values of power factor, prospective current and arcing angle (point-on-wave at which arcing starts) will coincide on any particular fault. The values plotted in Fig. 6.1 are thus inherently safe by an adequate practical margin.

There are other factors which tend to increase the safety margins still further. The pre-arcing I^2t value of the major fuse is taken from tests in which the fuse actually blows and cuts off in a few milliseconds. The total I^2t let-through of the minor fuse on the other hand includes both the pre-arcing and the arcing time, which in total are rather longer than the pre-arcing time at which the I^2t of the major fuse is computed. During this extra time there is heat loss from the major fuse element so that its temperature may remain substantially below the melting point. The effect is further accentuated by the temperature coefficient of resistance of the element. The heat generated in the fuse is proportional to I^2rt ($\int i^2 r dt$) where r is the resistance of

TABLE 6.2

Discrimination tests at a low fault current

Major fuse-link		Minor fuse-link		Prospective current (A)
Rating (A)	Test reference	Rating (A)	Test reference	
100	F	80	h	720
		80	i	
100	G*	80	j	
		80	k	

*Same major fuse-link for three successive test shots.

the fuse and is a variable due to the temperature coefficient. Since the temperature of the element in the major fuse does not rise to melting point its resistance does not rise in the same order as that of the minor fuse (which blows).

It is not easy to evaluate these effects because they vary with different designs. They are most evident however in those designs of element in which the metal has a relatively high melting-point. The elements of the fuses illustrated are of high-purity silver; this has the further advantage of retaining its form even up to temperatures approaching its melting-point. The factors which tend to assist discrimination are therefore favourable to this design.

Table 6.2 shows a programme of tests at the lower fault currents. Again the major fuse is 100 A, but in these tests the major and minor fuses may be much less than

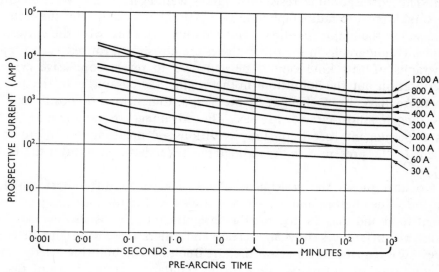

FIG. 6.5

Time/current characteristics. For pre-arcing and total operating I^2t values for times shorter than 0.02 sec, see Fig. 6.1

2:1 for faults at which the minor fuse does not exhibit cut-off. Major fuse (G) is made to withstand the let-through $i^2 t$ of three minor fuses (i, j and k) blown in succession. Fig. 6.4 is an oscillogram recorded from one of the tests shown in Table 6.2. The pre-arcing $I^2 t$ of the minor fuse-link is greater than would occur for the higher prospective currents because the pre-arcing time is appreciably longer and this allows more heat to be lost from the element before it melts. The arcing $I^2 t$, however, is small and the total $I^2 t$ is much less than the minimum pre-arcing $I^2 t$ of the 100 A fuse.

If there had been a 90 A fuse available in the range as a standard rating, there is no doubt that this too would have discriminated with the 100 A major fuse. This is because the longer pre-arcing times apply equally to both the major and minor fuse-links.

The practical effect of these results is that successive current ratings shown on a family of time/current curves as illustrated in Fig. 6.5 will discriminate satisfactorily. The curves show only those currents which correspond to pre-arcing times longer than about one cycle, i.e. currents below which the minor fuse-links begin to exhibit cut-off.

6.4 Physical examination of major fuse-links

The final proof of successful discrimination lies in the ability of the major fuse-links to resist any form of deterioration. Accordingly, all the major fuse-links involved in the tests and as listed in Tables 6.1 and 6.2 were thoroughly examined to detect any evidence of change in their physical states.

Prior to the tests, the resistance of each fuse-link had been measured and recorded. The corresponding resistance values were again measured after the tests. The fuse-links were also radiographed in two planes before and after the tests.

Fig 6.6 shows the major fuse-links photographed after test with the corresponding radiographs. (Radiographs in the other planes are not shown for reasons of space.)

Radiographs of this kind cannot be expected to reveal conclusive evidence of non-deterioration but the definition (in the original radiograph) is sufficient to show whether there have been changes in the shape, dimensions or positioning of the elements due to short-circuit stresses. The advantage of radiography is that it provides a permanent record of the conditions of the samples before they are dismantled. In the present case the radiographs taken before and after the tests were identical.

Fig. 6.6 also shows the resistance values. The small differences which occur between the values before and after test are well within the measuring errors of the instrument used and may be regarded as insignificant. This is also evident from the fact that some differences show an increase and others a decrease. The differences in the values between one fuse and another are due to manufacturing tolerances. These are representative of normal manufacturing standards for the type of fuse-link in question. The effects of differing resistance values were taken into account when assessing results to be sure that a minor fuse with a resistance value towards the

MAJOR 100 A FUSE-LINKS AFTER TEST	RADIOGRAPHS OF MAJOR 100 A FUSE LINKS AFTER TEST	RESISTANCE VALUES ($\mu\Omega$)	
		BEFORE TEST	AFTER TEST
	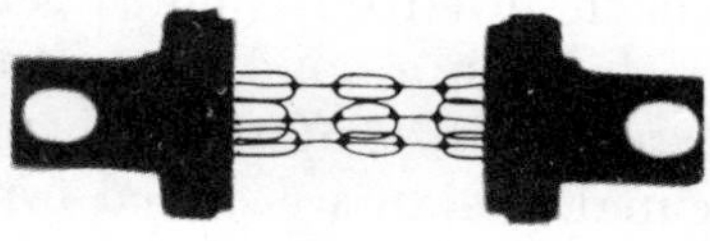	595	590
	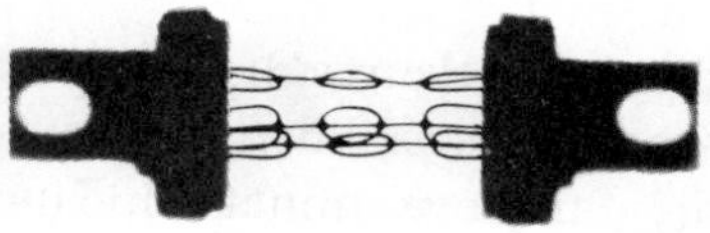	600	600
	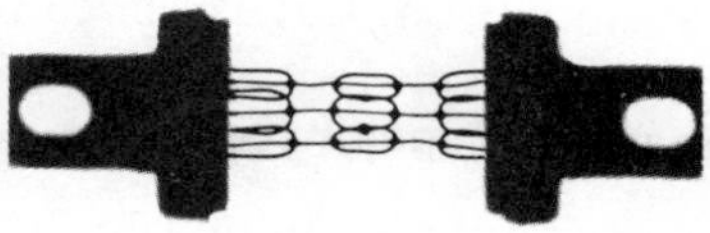	590	595
	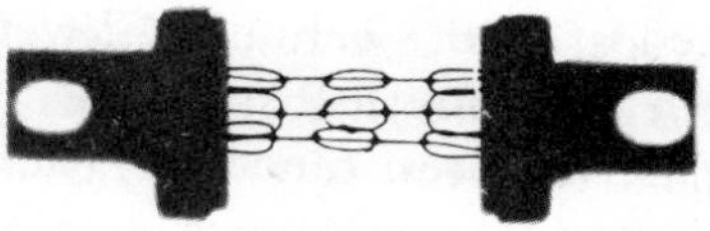	590	590
	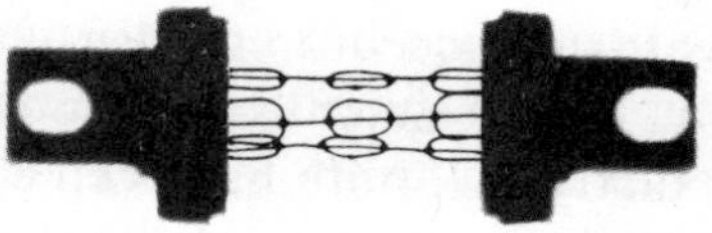	590	590
	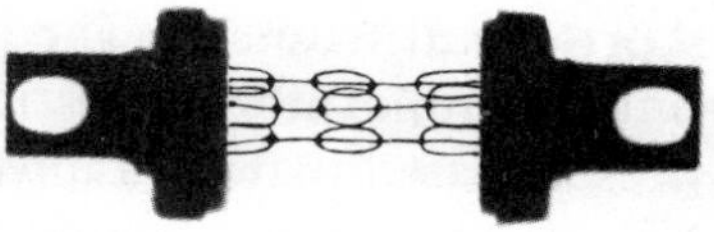	600	605
	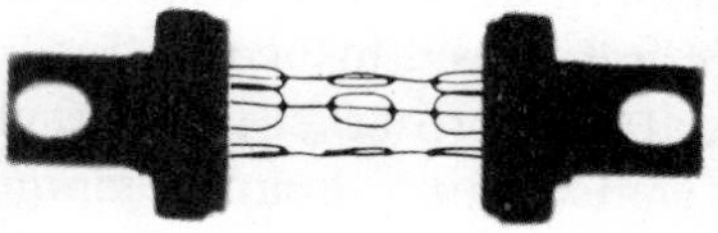	600	600

FIG. 6.6
Major fuse-links after tests

lower limit of tolerance would discriminate with a major fuse having a value towards the higher limit. Among the many tests carried out this circumstance did arise, with entirely satisfactory results on each occasion.

The fuse-links were then dismantled in stages with examination at each successive stage. Particular attention was given to the condition of the joints between the elements and the inner end-caps to ensure that no deterioration had taken place at this point, because such an effect could be serious. If, for instance, a fuse-link were to remain in service in such a condition, danger could arise when the next fault occurred because arcing might start at the weakened joint before it started at that part of the element which is designed for dealing with the arc. Arcing near the end-caps could result in a burning-through with emission of gas and flame and consequent flashover.

Finally the individual elements were removed and subjected to visual and microscopic examination. Fig 6.7 shows some of the elements and other components laid out during examination. The definition of the printed photograph is perhaps not sufficiently clear to show the evidence of non-deterioration which is the object of the examination, but it serves to illustrate the methods employed in the investigation. There was in fact no deterioration in these cases. The joints were metallurgically unchanged and the silver wire as bright and clean as new.

6.5. Tests on pre-heated fuse-links

The tests previously described were conducted to study the effects of different fault conditions. For this purpose the fuse-links were cold at the start of the tests. In service fuses may have been carrying varying degrees of load immediately before the fault occurs. Further tests were therefore carried out with the major fuse-links in the 'warm' or pre-heated condition, and the minor fuse-links cold. This is the most severe condition which can occur in service as regards temperature difference. It will be realized that if both major and minor fuse-links are warm their relative behaviour will be exactly as if both had been cold.

The same procedure as previously described was repeated. Fuse-links of 50 A and 100 A ratings were again tested at various fault levels and discrimination was achieved in each case. The difference in I^2t values obtained from the oscillograms did not appear to be significantly different from those obtained on the cold fuse-links. Such difference as there was, was not obvious within the limits of methods used for measuring the oscillograms because the current traces in these cases are rather small. A difference must exist, but it is of no practical significance.

Again the criterion was the condition of the major fuse-links after tests. Examination showed them to be entirely unchanged, which proved that the effects of pre-heating are well within the safety margins inherent in the published data. Fig. 6.8 is one of the several plates recorded to show the condition of the elements and other components. Again the printed definition is not good enough to show the essential details, but the photograph is included to illustrate some of the variations in test conditions which were considered.

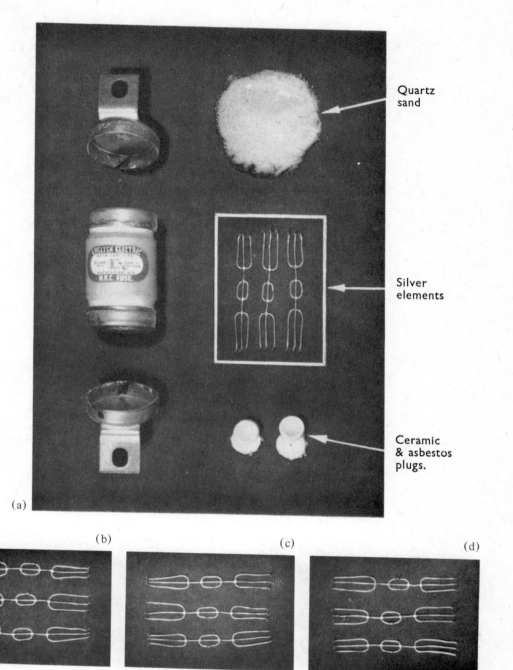

Quartz
sand

Silver
elements

Ceramic
& asbestos
plugs.

(a)

(b)   (c)   (d)

FIG. 6.7
Samples of elements from 100 A major fuse-links after discrimination tests (fuse-links starting from cold condition): (a) sample dismantled after test to show components; (b) after three tests at 46 kA in series with a 50 A fuse-link; (c) after test at 1.8 kA in series with a 50 A fuse-link; (d) after three tests at 0.72 kA in series with an 80 A fuse-link.

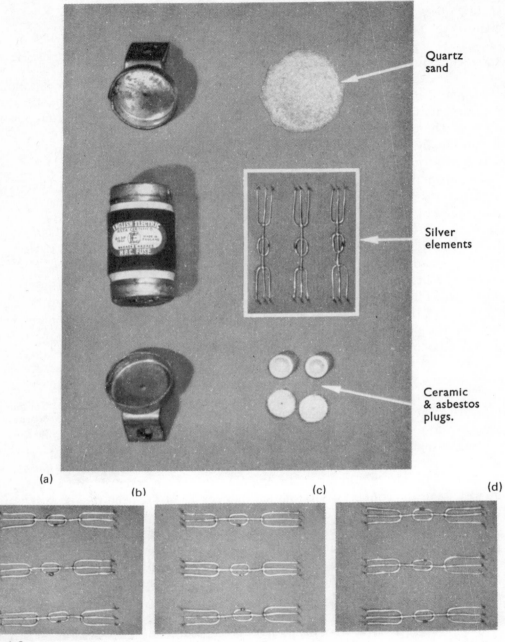

Quartz sand

Silver elements

Ceramic & asbestos plugs.

(a)

(b)   (c)   (d)

FIG. 6.8

Samples of elements from 100 A major fuse-links after discriminating with 80 A minor fuse-links at 300 A fault current (all major fuse-links carried stated load for 90 min before test): (a) sample dismantled after test to show components; (b) after test with one minor fuse-link, major fuse-link pre-heated to 66% full load current; (c) after test with one minor fuse-link, major fuse-link pre-heated to 100% full load current; (d) after test with two minor fuse-links, major fuse-link pre-heated to 66% full load current

6.6 Discrimination in practice

H.R.C. fuses are in widespread use throughout the world operating under all conditions. They are handled for the most part by working electricians, many of whom are employed for their practical, rather than their technical or academic, knowledge. For them it is necessary to reduce technical data to bare essentials and in some cases to simple 'rules of thumb'. This does not mean that the methods of tests, and other information upon which the data is based, can be cursory. On the contrary, it is necessary to be all the more careful to ensure that nothing is left to chance. The tests described in this chapter are an instance of the sort of work which is continually being carried out to ensure that familiar and established practices are sound and founded on technical truths.

The attractiveness of the H.R.C. fuse is its simplicity of application. The fact that its functions are complex does not need to worry its users unnecessarily, provided that they can trust its performance and be sure that they have sufficient information in a concise form to use it to good effect. The manufacturer's responsibility is to ensure that it is rated properly and that it obeys its designer's intentions implicitly. If these conditions are met, its behaviour can be stated in fairly straightforward terms. For instance, a 'rule of thumb' ratio of 2:1 between ratings of the fuse-links illustrated will suffice to give discrimination under all normal circumstances. If a closer ratio than this occurs and cannot be remedied by rearrangement of the circuits, the prospects of discrimination may be judged by reference to the fuse data provided in cases where the fault level can be estimated to a reasonable degree.

Such rules will, of course, only apply when the fuse-links are manufactured to the same accuracy as those illustrated in the tests described. These are manufactured within a tolerance of ±5 per cent of a mean resistance figure and this is ensured by measuring the resistance of every fuse made, as well as by close attention to all other aspects of quality control. Few protective devices in common use on medium-voltage systems can provide such a high degree of accuracy as the modern H.R.C. fuse and none can do this so consistently throughout a life of perhaps twenty years or more without re-calibration or maintenance. Non-deterioration is particularly important in the context of discrimination because of the need to ensure stability of characteristics. It thus governs the validity of characteristic curves, because these usually refer to the performance of the fuse (or other device) when new.

It has been shown that it is a relatively simple matter to choose H.R.C. fuses to achieve discrimination notwithstanding the complexity of their function. There is no fundamentally new idea in fuse discrimination. The tests described are no more than proof that fuse-links can be manufactured (and indeed have been available for many years) which discriminate with predictable accuracy.

It is, however, true that planning engineers concerned with the calculation of protection problems or with the design of equipment in which H.R.C. fuses are incorporated are continually demanding greater precision in the definition of fuse parameters. This is because the ratings of cables, busbars, motors and other electrical equipment are constantly tending to increase. Circuit conditions also tend to become more severe and all these factors intensify the problems of protection. Researches such as those described therefore have further value in that they point the way to a more scientific approach to fuse applications where this is needed. All protective devices are equally affected by this pressure of events and it is reassuring to be able to show that the H.R.C. fuse is well placed to maintain its position in the fields in which it is already established and in the new applications which are constantly appearing. During the past few years the rupturing capacities of 'English Electric' H.R.C. fuses have increased in advance of requirements: accuracies of characteristics are adequate for the foreseeable future and non-deterioration provides the stability and reliability which modern conditions demand.

7 H.R.C. fuse protection of high-voltage power circuits

A number of accidents resulting from the electrical failure of the terminal boxes attached to large high-voltage motors had focused attention in the early 1960s to the hazards associated with these situations. Various attempts were made to produce terminal boxes with pressure relief to dissipate the arc products from short-circuit faults but the energy limiting properties of H.R.C. fuses were shown to provide a better alternative. In the 1970s the use of fuses for this purpose has become an accepted practice but at the time this paper was written incontrovertible proof was demanded to show that fuses would perform adequately in this respect.

A series of high-power short-circuit tests were arranged to be witnessed by an independent symposium of senior engineers from all branches of industry in the United Kingdom. The programme was duly carried out under the scrutiny of the assembled experts who satisfied themselves as to the validity of the tests and full implications of the results obtained.

The tests left no doubts concerning the ability of the fuse to limit short-circuit damage. The problem which attracted debate was that of how to select fuses for particular motor duties.

New developments which have now emerged from these events have happily coincided with the introduction of vacuum interrupting devices. The combination of vacuum interrupters and fuse now meets most of the problems which are arising from the greatly increased use of high-voltage motors, and the previous vulnerability of terminal and cable box connections is no longer considered to be a serious problem.

Originally published in *Electronics & Power*, April 1964.

7.1 Introduction

The ability of H.R.C. fuses to limit fault energy in medium-voltage circuits is well known. It assumes added significance when applied in the protection of high-voltage systems. This is not because the higher voltage in itself will necessarily increase the fault incidence but because of the greater potential damage and devastation whenever a fault occurs. Incipient faults develop more readily, and power arcs are more easily sustained at a high voltage than at a low one. As many, and perhaps the majority, of faults that occur in practice are due to arcing, the H.R.C. fuse is the natural remedy for keeping damage to a minimum.

It is not only the damage to the equipment directly involved in the fault that needs to be considered but also the consequential hazards, such as danger to

a

b

FIG. 7.1
*Terminal box and fuses used in tests: (a) method of
mounting; (b) terminal box with cover removed, showing
new condition of interior and method of creating
three-phase short circuit*

personnel and fire risk. There is also the question of the time taken to repair the fault damage and to reinstate electrical services. In the case of auxiliary services to power stations, ships or large industrial installations, these may well be the major factors.

7.2 Increase in fault levels

In recent times, attention has been focused on these problems by adverse experience with faults in high-voltage motor terminal boxes, particularly at 3.3 kV. Investigations into the reasons for the faults and an evaluation of the hazards involved indicate that the main reason for the increasing troubles has been the emergence of higher fault levels, which have grown from something less than 100 MVA at 3.3 kV to 150 MVA and even approaching 250 MVA within the last decade.

Another reason is that a much greater number of 3.3 kV motors are being used now than formerly. There are also greater concentrations of motors in a given location. Fault damage associated with motors is liable to be more serious than that with other classes of equipment, because motors are usually situated in places easily accessible to personnel.

It is admitted that new measures are necessary, and a great deal of attention has been paid to the finding of possible remedies. Steps have been taken to improve the design of motor terminal boxes, but other parts of the circuit that are equally vulnerable do not show the same possibility of improvement. This is why the H.R.C. fuse holds the key to the situation. It alleviates hazards anywhere in the system and reduces damage to easily manageable proportions.

A comprehensive series of tests was recently carried out before representatives of the electrical industry, to demonstrate the abilities of H.R.C. fuses in this role. The purpose of this chapter is to describe the tests, to explain the basis on which they were planned, and to outline the conclusions that may be drawn from them.

7.3 Demonstration tests

The tests were carried out to demonstrate that the energy-limiting characteristics of 3.3 kV H.R.C. fuses could effectively minimize fault damage in motor terminal boxes and cable boxes. The object of the demonstration was to prove that the energy let through by the fuses into the fault arc could be contained within the terminal or cable box so as to avoid fracture of the box or emission of arc products that might be dangerous to personnel or adjacent equipment.

The boxes used, shown in Fig. 7.1 (a), were of a standard approved design in general use. Both were fabricated in sheet steel, the terminal boxes having pressure-release apertures covered by thin metal diaphragms and the cable boxes being compound filled. The fuses used were from a standard range of known performance, tested to 250 MVA and A.S.T.A. certified to B.S. 2692:1956.

a

b

FIG. 7.2
*Terminal box subjected to 250 MVA fault with fuse
protection: (a) terminal box with cover removed to show
condition of interior (no external emission of arc products
occurred during test); (b) close-up view of condition of
terminals with rubber shrouds partially removed*

The test conditions were chosen to produce the most onerous combination of
circumstances that could ever be envisaged as occurring in service, with reference to
the severity of test circuit, the nature of the fault and the choice of fuse rating.

7.4 Test circuit

The test circuit was fed from a high-power supply capable of delivering 250 MVA at
full voltage, the power factor was of low order and the circuit made by random
switching.

Tests were conducted at a variety of fault levels (up to 250 MVA at 3.3 kV). The maximum short-circuit stress does not necessarily occur at the highest fault level when fuses are used. This is a function of fuse design and performance. Hence it is necessary to carry out tests at intermediate fault levels to produce the most onerous conditions that can occur in service relative to the fuse performance and so demonstrate that the damage within the terminal box, even in these conditions, is reduced to manageable proportions.

By selecting a fuse of large current rating, the tests may be taken as valid for all lesser ratings in the same range. The let-through I^2t of a fuse, and thus the damage at the seat of the fault protected by the fuse, varies almost directly as the square of the current rating. This rule is independent of the fault level and means that the fault damage in the smaller circuits can be reduced to very small proportions. The actual values to be expected can easily be derived from the fuse-performance data, which are available for the whole range and which have been obtained from actual fuse tests.

Another point to be noted is that the 400 A fuse-links used in the tests were composite fuses consisting of two cartridges in parallel. The principle of operating fuses in parallel, subject to known safeguards, has been amply proved elsewhere. The present tests provided a further opportunity to prove this principle in specific operating conditions and to show that there is no upper limit, within practical terms, to the current rating of fuses.

7.5 Faults

Faults in service are fortuitous, and the degree of damage depends on a variety of factors; e.g. a simple bolted fault produces no significant damage at the seat of the fault itself, whereas a high-impedance fault limits the damage, because the current itself is limited and the inductive energy stored in the circuit at a minimum.

Since the present tests were intended to investigate the maximum damaging effect of the fault, considerable study was required to create fault conditions that were representative of the worst that could occur in service and that could be consistently repeated.

Precipitation of a fault and maintenance of a power arc are largely dependent on the system voltage. 3.3 kV will easily maintain an arc within the confines of a motor terminal box or a cable box once there has been sufficient ionization to allow the arc to strike. Thus, the most convenient way of creating a fault for test purposes is to bridge the terminals or conductors with open fuse wire, carefully chosen to give optimum results, as shown in Fig. 7.1 (b).

For the particular test arrangements under consideration, it was found that a single strand of 22 s.w.g. tinned copper wire produced sufficient ionization in relation to the volume of the enclosures to allow a power arc to develop instantaneously. A condition could therefore be obtained in which maximum arc damage relative to the 400 A fuse could be produced at the seat of the fault. This was independent of the fuse wire after the arc had been struck and did not either assist or retard the normal operation of the H.R.C. fuse. Numerous test shots carried out prior to the formal test series had shown that this method produced entirely consistent results. This was proved by the test oscillograms of successive shots, which were similar for given circuit conditions.

For the cable box, the fault was again initiated by a single strand of 22 s.w.g. wire connected between the cable cores near the crutch and under the compound.

7.6 Test report

Following is an extract from the test report issued in respect of the test mentioned:

Three-phase, short-circuit tests were made to show that high-voltage fuses could be used to protect motor terminal boxes from internal arcing faults to the extent that no dangerous arc products would be emitted. The terminal boxes used were to a standard approved design in general use.

Tests were made on four terminal boxes, two with faults at the terminals and two with faults between cable cores near to the cable crutch under the compound within the cable box.

The first of these, with the fault at the terminals, was tested at three values of prospective current, namely 10.5, 22.3, 39.6 kA. In all these tests the circuit was protected by fuses, and no disturbance appeared external to the box during the fault occurrence. The fuses operated in a satisfactory manner without demonstration.

The second box, again with the fault at the terminals, was tested at 39.7 kA without fuse protection. The test resulted in emission of considerable flame and gas from the terminal box.

The third box was arranged with a fault between the cable cores within the cable box to simulate a cable-crutch fault. The test was at 40 kA with fuse protection. No external disturbance occurred from either the terminal box or cable box. The fuses operated in a satisfactory manner.

Finally, the fourth box was arranged as for the third test, except that no fuse protection was provided and the test current was 39.8 kA. The test resulted in a violent explosion, in which considerable quantities of flame, gas and molten compound were emitted. The cover of the terminal box was blown off with considerable force to a distance exceeding 20 yd.

A summary of the test results is given in Table 7.1

7.7 Conclusion from tests

The tests showed that the performance of the fuse was exemplary. This was to be expected, since the fuses used were taken from a range that was already tested to 250 MVA and A.S.T.A.-certified for compliance with B.S. 2692, and the British

a

b

FIG. 7.3
*Motor terminal box subjected to 250 MVA three-phase
short circuit without fuse protection: (a) terminal box with
cover removed, showing condition of interior after test;
(b) cinéstill taken at time of fault (pressure-release vent in
operation)*

Standard and A.S.T.A. rules for testing are even more severe than the three-phase tests described above.

No external or visual effects or noise, from either the fuses or the fault, were evident during tests in which the fuses interrupted the circuit. The damage sustained was so slight (see Figs.7.2(a) and (b)) that, had it been in actual service, the equipment would have been fit for continued duty after only superficial cleaning inside the terminal box.

FIG. 7.4
Cinéstill of 250 MVA three-phase short circuit occurring in cable box without fuse protection

The damage sustained in the tests when fuses were not used and when the fault was interrupted after 0.2 sec rendered the box completely unusable (Fig. 7.3(a)). The violent emission of hot gases and flame from the fault (Fig. 7.3(b)) constituted a positive fire hazard and was such that it would have been a danger to any person who might have been in the immediate vicinity. Another hazard that is sometimes overlooked in test records but which was evident to those witnessing the tests is that of noise. The loud report that accompanies any fault that is unprotected by fuses is frightening and must be of vital concern to those who are responsible for the safety of personnel.

The results of the tests on the cable boxes were even more dramatic (Fig. 7.4). The tests with fuses did no more than scatter melted compound upwards into the terminal box, but there were no significant external effects. Tests without fuses produced a spectacular and violent explosion accompanied by flames and molten materials that were shot out, together with pieces of steel from the assembly, over considerable distances. The noise was similar to that produced by a high-explosive charge, and it has been estimated that the total power in the fault and the rate at which it was released were equivalent to an explosion of several handgrenades.

7.8 Service requirements

The problems of fault protection and the remedies suggested are common to all voltage levels. The demand for larger motors and the increasing tendency towards higher voltages have become more evident during the last few years. In the U.K., 3.3 kV motors are widely used, and 6.6 and 11 kV motors are not uncommon; elsewhere in the world 6 kV is common for motors of even relatively small horsepower rating. In all these cases the H.R.C. fuse finds application. A good deal

TABLE 7.1
Summary of test results

Motor terminal boxes and cable boxes protected by 400 A, 3·3 kV 250 MVA H.R.C. fuses

Test	Severity of test circuit				Results of tests			
	Voltage, r.m.s. (kV)	Prospective current (kA) Peak	r.m.s.	Power factor	Cut-off current, peak (kA)	Operating time (msec)	Arc voltage (kV)	Remarks
Terminal box protected by fuse (10 kA)	3.42	(R) 17.5	10.2		(R) 13.9	16.0	2.63	No visible disturbance; performance satisfactory; diaphragm slightly bowed
		(Y) 23.5	10.3	0.11	(Y) 13.5	13.8	6.0	
		(B) 22.0	11.0		(B) 17.8	16.0	6.0	
Terminal box protected by fuse (20 kA)	3.43	(R) 37.6	21.8		(R) 17.2	11.6	4.9	No visible disturbance; performance satisfactory
		(Y) 47.8	22.0	0.14	(Y) 10.7	11.6	2.92	
		(B) 45.4	23.0		(B) 25.4	8.9	7.3	
Terminal box protected by fuse (40 kA)	3.41	(R) 68.1	39.7		(R) 24.2	4.0	5.0	No visible disturbance; performance satisfactory; slight burning of terminals; slight deposit of brown soot
		(Y) 83.3	39.8	0.12	(Y) 11.7	8.3	2.87	
		(B) 73.6	39.2		(B) 26.2	8.3	6.0	
Terminal box without fuse (40 kA)	3.42	(R) —	39.8				n.a.	Excessive flame, gas and noise
		(Y) —	39.9	0.12		210		
		(B) —	39.3					
Cable box protected by fuse (40 kA)	3.3	(R) —	40.2		(R) 22.7	4.4	n.a.	No visible disturbance; performance satisfactory; compound blown into terminal box
		(Y) —	40.3	0.12	(Y) 7.73	4.4		
		(B) —	39.6		(B) 22.9			
Cable box without fuse (40 kA)	3.43	(R) —	39.9				n.a.	Excessive flame, gas, sparks and noise; parts blown out for distances exceeding 20 yd
		(Y) —	40.0	0.12		225		
		(B) —	39.4					

(R), (Y) and (B) refer to the red, yellow and blue phases.
n.a. = not available.

of experience has already been gained in this connection, and there is no lack of fuse data on which to proceed with the applications now emerging.

The choice of fuse for any particular application is not difficult to make, provided that certain fundamental requirements are borne in mind. Where H.R.C. fuses are used in motor circuits, their first duty is to provide short-circuit protection and to minimize fault damage. They must therefore be properly co-ordinated with associated protective devices to achieve this objective and must be chosen so that they do not change or tend to operate except when a short circuit occurs. Essentially, this means that a relationship must be established between the time/current characteristics of the fuse and the operational requirements of the motor to ensure that the fuse retains its essential characteristics during the normal life of the equipment in which it is mounted.

The position of H.R.C. fuses in the circuit and the problems of mounting them vary with circumstances. For motor circuits, it is an obvious advantage to include the distributor cable within the zone protected by the fuses, because of the economies accruing from the reduction in short-circuit stress. The same reasoning applies also to other components in the circuit, and these should be included in the protected zone where practicable.

In most cases fuses are mounted as an integral part of switchgear equipment, and in these cases the physical arrangement becomes the problem of the switchgear designer. This usually simplifies the problems of co-ordination, but in some cases calls for ingenuity in reconciling the fuse position with the procedures of isolation and earthing.

8 H.R.C. Fuses in the protection of cables

The repercussions from the controversies which arose in connection with the protection of P.V.C. insulated cables led inevitably to a reappraisal of cable protection in general.

The protection of cables against low overcurrent damage had been successfully achieved over many years by the use of H.R.C. fuses but had been taken for granted because the service record was virtually unblemished. This aspect of protection is incidental to the short-circuit protection for which fuses are usually required and was never adequately defined in British Wiring Regulations or Codes of Practice. A demand now arose for formal recognition of the problem. U.S. and Canadian practice had prescribed low overcurrent protection for cables over many years but whether this had produced better results than the apparent laissez-faire British attitude is open to question. Nevertheless clarification and assurance on the point was overdue.

Investigations confirmed that H.R.C. fuses provide an adequate degree of overload protection and explained the technical factors involved. The way was then clear to frame new regulations as and when required.

Originally published in *Electrical Times,* January 1965.

8.1 Introduction

Medium-voltage cables have long enjoyed a high reputation for reliability in this country. Apart from questions of quality in manufacture and proper choice of materials, this stems from the fact that they have been generously rated both dielectrically and (in the past) thermally. It also suggests that the protective devices used for cable protection have been satisfactory and effective.

Because experience with cables has been so favourable it follows that the subject of protection has not in the past been a particularly controversial issue. But recent changes in the conceptions of rating for some types of cables due to the introduction of P.V.C. thermoplastic insulation have brought the problems of protection into new focus and have necessitated a re-examination of existing protective devices and, in particular, of fuses. It is evident from even a cursory examination, that the capabilities of contemporary H.R.C. fuses have in some respects not only been taken for granted—their performance has been hidden under the bushel of out-of-date specifications and regulations.

The H.R.C. fuse has been the most commonly used protective device for cables in industrial installations and public supply networks for many years: it has also been widely used for protecting small wiring circuits and domestic installations. In the vast majority of all these cases it is the only form of protection which has been provided. Service experience has proved that fuses of this kind have given satisfactory protection against all values of overcurrents from the overload zones up to the highest short-circuit levels.

The ability of the H.R.C. fuse to protect against the lower orders of overcurrent has not received so much prominence as its ability to deal with the higher orders of short-circuit currents because it is the latter which has distinguished it from other less versatile devices. Service experience has proved it to be satisfactory for the overload protection of cables, and events are now making it necessary to acknowledge this function with greater emphasis.

8.2 Cable protection

In recent years the introduction of synthetic materials, including thermoplastics, has permitted higher continuous operating temperatures and, therefore, higher current ratings; but some materials, such as P.V.C., are more sensitive to overload conditions. At the same time the greatly increased quantities of cables installed and the dependence of industry upon the service they give, has led to re-evaluations of the safety and actuarial aspects of rating and protection. Opinions have differed concerning the requirements of overcurrent protection, but such differences are relatively easy to resolve. The main difficulties seem to be those which have arisen due to standards, specifications and regulations (relating to cables and protective devices) having tended to lag behind the requirements of practice. Generally speaking, there are no insuperable difficulties in meeting the technical problems arising in practice, but it has not always been possible to reconcile standards and regulations with the actual circumstances of a particular case, to achieve the most economical solution.

A recent example in this context was the problem which arose following the introduction of armoured P.V.C. insulated cables. The ratings of these cables were conditional upon the form of protection provided. In this case both the specifications for fuses (e.g. B.S. 88) and the regulations governing their use (e.g. I.E.E. Wiring Regulations) were couched in terms which were not entirely compatible with the new conceptions of cable rating. The misunderstandings which arose from this situation had to be tackled energetically to avoid difficulties, and changes have had to be introduced to remedy the position. The basis of ratings for armoured P.V.C. insulated cables has been clarified in an amended publication issued by the Electrical Research Association (Amended Supplement to Reference F/T. 183) and amendments to fuse specifications have now been proposed which conform to this basis. Corresponding amendments to regulations governing the installation of both fuses and cables may also have to be considered. *The main point to be appreciated is that no physical modifications have been necessary in the actual fuses or cables. The only changes necessary have been to the written word.*.

The airing of views on the problems associated with plastics-insulated cables has inevitably caused the whole question of cable protection to be reviewed. Existing methods hallowed by long experience are being subjected to technical scrutiny. The need for additional information concerning the performance of both cables and fuses has become evident. This does not in itself condemn existing methods, nor does it introduce any new principles. Fundamentally, it is an attempt to find a more scientific basis to describe the best and safest practice already evolved by experience. It is yet another manifestation of the increasing importance of electricity as a service and of the demand for higher standards of safety. This chapter is an attempt to throw light on one aspect of the problem, by presenting empirical evidence of fuse performance in relation to the overload protection of cables in general. Multi-core P.V.C.-insulated cables as such have been dealt with in the chapter entitled *Protecting Armoured P.V.C. Cables.*

8.3 Parameters of protection

Cables need to be protected against all overcurrents from relatively small overloads to the highest short-circuit currents. The H.R.C. fuse is able to provide short-circuit protection up to the highest values of fault current and, in addition, so to limit the fault energy as to keep fault damage to a minimum. These capabilities are well documented and, generally speaking, the higher the overcurrent, the easier it is for a fuse to deal with it. It is in the overload zones that the fuse characteristics require to be further explained because this aspect of fuse performance has hitherto gone almost unrecognized.

The minimum fusing current of a fuse (which may be approximately compared to the minimum setting of a circuit-breaker) determines the degree of sensitivity to the smaller overloads. No protective device can be expected to run continuously on all values of current above its own rating. Similarly, some margin must be allowed between the cable rating and the minimum operating value of the device protecting it, if the cable is to be fully utilized. Too small a margin would result in unnecessarily early disconnection of the cable, which could lead to consequences more serious than those resulting from an unduly high margin. Some small degree of risk has to be accepted, and the optimum magnitude of margin is related to this. In practice, the hazards are very small, because ratings assigned to cables provide for some overloads; and because overloads when they occur are in most cases transitory.

Where overloads persist to the point at which the cable fails, the resulting fault usually causes the protective device to operate very quickly. Thus, even in the small minority of cases in which the cable itself may suffer damage, danger in the general sense is unlikely to result.

Where persistent overloads are known to be a possibility, and are likely to lead to serious consequences, the accepted precaution is to install cables of adequate rating in the first place. The economic incentives for this policy are obvious. From the point of view of the H.R.C. fuse there is another reason why the hazards resulting from small overloads are small. It is that fuses, being thermal devices, are self-adjusting in their characteristics towards safety.

The supply industry provides a good proof of these contentions. The day-to-day loading of supply networks is decided by consumers, and in the short term is beyond the control of the supply authority. Thus, distributor cables are inevitably overloaded from time to time. Nevertheless, H.R.C fuses are used exclusively for the protection of such cables, and in most cases the ratings of the fuses used are well in excess of those of the cables. It is seldom that a cable suffers damage to the extent that it needs to be replaced. Faults due to causes other than overload do of course occur, but these are almost always localized and occur at joints or other places where the cable has been disturbed. A fault due to deterioration in the run of the cable itself is a rare occurrence.

For overcurrents which come within the sensitivity of the H.R.C. fuse there is little difficulty in ensuring that the time/current characteristics of the fuse come within the thermal withstand of the cable (expressed in terms of conductor temperature and time). The time/current withstand of cables varies with circumstances, and appropriate correction factors may be required for different situations, but a purely nominal value of short-time rating of the cable is sufficient to allow a suitable choice of fuse to be made. In a recent document relating to P.V.C.-insulated cables it is laid down that these cables when installed in air are able to withstand currents not exceeding 150 per cent of their ratings for four hours. Most other forms of cable in common use have a higher withstand than this, which implies that a device which will protect P.V.C. cables can be reasonably expected to be satisfactory for other types.

The factors involved in the determination of the thermal withstand of cables are in fact rather complicated, but for practical purposes they have to be put into reasonable perspective and resolved into working rules. These involve the use of corrective factors which accompany tables of rating. It has never been considered necessary to apply similar corrective factors to fuses except, perhaps, in respect of wide variations of ambient temperature. The higher temperature within fuse enclosures which would otherwise effect the fuse time/current characteristics are usually offset by diversity and load factors and it is only in severe conditions that correction needs to be made to the fuse rating on this account.

The choice of fuse characteristics for particular applications has to a large extent been evolved empirically and by experience. Cable protection, which represents only one of several duties to be considered, has not hitherto been the most critical. Whether it will become necessary to match fuse characteristics more closely to cable characteristics in the future, depends entirely upon how closely to their working limits cables and other protected equipment are to be rated. Present fuse characteristics are easily capable of catering for reasonable changes in cable use, and there are no inherent limitations in fuse design which would prevent them following any further trends in cable practice towards less conservative ratings. In the ultimate case the H.R.C. fuse, connected to a grossly overloaded cable of smaller rating than itself, is usually in a position where it is affected by the higher temperature of the cable and will blow at a lower current than it otherwise would have done under more normal conditions.

8.4 Fuse behaviour

Fuse protection in respect of overload currents depends upon the time/current characteristics which are published by the fuse manufacturer in the form of the familiar time/current curves. At the relatively longer times corresponding to the minimum currents at which the fuse will operate, these curves tend to become asymptotic. However, they can never become completely so and any fuse subjected to a continuous current equal to or even slightly below its declared minimum fusing current must eventually blow. When the fuse is pushed to its limit for prolonged times it will invoke its own safety factor and open the circuit. The times required to cause operation in this manner are longer than could reasonably be included in published time/current curves, but experience indicates that they are nevertheless usually within the survival time of associated cables. This may imply that conventional curves do not entirely reveal the ability of a fuse to protect a cable under extreme overload conditions.

It should also be remembered that published time/current curves are derived from tests under prescribed conditions and that service conditions often differ from these. The errors due to such differences do not cause difficulty in the vast majority of cases, because they are biased in a safe direction. Test conditions in this context are prescribed so as to be representative of the worst conditions (for the fuse) which can occur in normal service, i.e. with the fuse mounted in a reasonably open position at an ambient not exceeding 25°C and with the fuse-link at room temperature at the commencement of the test. In service the fuse-link is likely to be in an enclosure and to be warm from having carried load prior to the fault current occurring. Such conditions would cause it to blow rather more quickly at a given current than indicated by the published curve.

8.5 Minimum fusing current

Minimum fusing current is a value corresponding to operation in an arbitrary time obtained under prescribed test conditions. Alternatively, it is a value of current corresponding to a chosen value of time indicated on a time/current curve which is itself obtained from prescribed testing conditions. By convention in British Standards, fuses are classified in terms of 'fusing factor' which is a name given to the ratio (minimum fusing current/fuse rating). Fusing factor is subject to the same variations as the time/current curves and by the same rule is also subject to the same safety factors. The arbitrary times chosen for the determination of fusing factor are not dissimilar to the values of time normally associated with the overload withstand of the cables. For example, a fuse having a fusing factor of 1.5 is adequate for protecting a P.V.C. armoured cable which is declared to be capable of withstanding overloads 150 per cent of its rating for four hours. The justification for this contention is that the short time rating for the cable, being the 'no-damage' limit, allows reasonable contingency, while the fusing factor of the fuse also includes the safety factors described.

The margin thus allowed between the withstand of the cable and the time/current characteristic of the fuse is ample to account for the manufacturing tolerances of both items, provided that these are of reputable quality. H.R.C. fuses manufactured to proper standards of quality can achieve an accuracy of ±5 per cent (in terms of current for a given time), and compare very favourably with the accuracy of alternative devices.

In distribution circuits, the I.E.E. Wiring Regulations prescribe that the rating of a fuse must not exceed the rating of a cable and where such regulations apply and where the fusing factor of a fuse is of the order of 1.5/1.6, overload protection of the cables is no problem. In motor circuits the rating of the fuse may be up to twice the rating of the cable which supplies the motor, it being assumed that close overload protection of the cable will be provided by the same device which protects the motor. Numerous instances occur in service in which the motor overload devices have become inoperative. Even in these instances it can be shown, and it has happened many times in service, that H.R.C. fuses have satisfactorily protected the cable. This is evidence enough that such fuses are not only adequate for distributor cables but have an important function as back-up protection for the cables in motor and other similar circuits.

8.6 Test evidence

That the H.R.C. fuse will protect cables against damaging overload currents is relatively easy to prove, as can be seen from the test results reported in this chapter. These have been chosen to represent a variety of cables under different overload conditions. Tests have been carried out on a number of current ratings with a variety of types of insulation, both single and multi-core, installed in air and bunched in conduits or other enclosures and with varying degrees of mechanical and thermal stresses. In each test overload currents were applied in ascending order with conductor and temperature measurements of the cable cores and connections were continuously recorded. At each current step, conductor temperature was allowed to reach equilibrium, and then the current was increased again. This sequence was continued until the fuse blew. Visual and radiographic examinations were made of all cable samples before each test.

The criteria upon which a fuse could be said to have successfully protected a cable were (a) the published limit of temperature for the particular cable, and (b) carefully evaluated evidence that the cable had suffered no detectable change or damage. The examination of cables in respect of the latter condition included visual and mechanical assessment of the insulating materials; radiographic and physical checking of the dimensions; insulation tests (5 kV for 15 min and after water immersion for 12 h, etc.); chemical and other checks as advised by the manufacturers of particular cable sizes.

At the conclusion of all the tests described no significant change in the cables could be detected.

Without exception, all the tests showed that the fuses protected the cables with a very considerable margin, so much so that although the evidence is entirely subjective, it is more than sufficient to support an objective verdict. The lower values of overcurrent are rather more severe than the higher values, but there is a decided increase in the margin of protection for relatively small increases in the value of overcurrent. This indicates that even in those cases where the fuse rating may be considerably higher than the cable rating, and where protection may seem to be doubtful, the hazard is confined to a very small band of overcurrent.

8.7 Other factors

The attention being focused on the overload protection of cables may tend to overemphasize this problem to the exclusion of other factors which may be more significant. Generally speaking, the likelihood of damage or danger through over-loading is small compared to other risks, as for instance, direct mechanical damage or other environmental hazards. Also, from an operational point of view, it has to be considered that the danger from an outage of a main supply may be more serious than the outage of a sub-circuit. If, for instance, the lights in a factory failed while machines were still running, the actual danger to personnel would outweigh any anticipated possibility of damage to a cable. Thus, the problems of discrimination and of co-ordinating the devices protecting cables with other devices in the system is often a more vital factor than overload risk. The provision of visual indication of overcurrents (such as ammeters) is in any case usually sufficient to prevent excesses of overloading.

Overall, it would appear that the hazards to cables are statistically few and those due to overloading are only a small proportion of the total. Moreover, a distinction must be drawn between those measures which may be required to prevent damage to the cable and those which are intended to prevent danger; damage to a cable itself may in many cases be of small account; danger is always serious but can usually be avoided by the judicious choice of electrical protective devices combined with the proper siting of the cable.

8.8 Rating philosophy

British and Commonwealth practice relating to the co-ordination of cable ratings and the protection provided for them, maintains as high a standard as anywhere in the world, and allows a more substantial margin of safety than in many other countries. This is sometimes taken as the incentive to search for more economical methods of utilizing cables and may in some cases be a valid motive. It is necessary to remember, however, that any change in the basis of rating of cables towards higher factors of utilization will obviously depend to an important extent upon the means by which the cable can be protected. H.R.C. fuses have so far proved themselves suitable for overcurrent protection at all levels. It is evident from the large margin shown in the test results described that there is still scope for them to continue to do so and to cater for possible re-rating of cables in the future. There is, moreover, scope for changes and improvements in the design of the fuses themselves. This is demonstrated by the fact that fuses of suitably designed characteristics are now used for the protection of equipment which is very much more critical than cables are ever likely to be.

8.9 Standards and regulations

The extent to which improvements in design, or even the known capabilities of existing fuses, can be utilized, depends to a large extent upon the sanction of standards and regulations. The indications are that these are tending to become more and more important, both nationally and internationally. In general, this tendency must be regarded as being both necessary and desirable inasmuch as it encourages proper standards of practice and of safety, but the responsibility for ensuring that

FIG. 8.1
Test rig for cables bunched in conduit: A, supply ends; B, fuse box;
C, cables under test; D, cable ends; E, weights; F, temperature recorder;
G, thermo couples

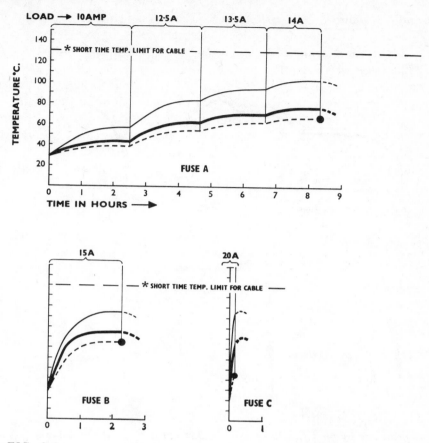

FIG. 8.2
8 A cables protected by 10 A fuse-links. Ten 1/.044 in P.V.C. insulated cables in conduit protected by 'English Electric' NITIO fuse-links. Five fuses in enclosure, fusing factor of single fuse in air approx 1.6

regulations do not stifle ideas or restrict development increases in direct proportion to the importance of the authority vested in the regulations. Also, great care needs to be taken to ensure that the specification of equipment for one purpose does not restrict or proscribe its application for another. It is particularly necessary to ensure that the specifications for related equipment are properly co-ordinated and this is admittedly difficult while individual items progress at different rates. In this context there is still a good deal to be done to bring H.R.C. fuse specifications up to date and a considerable amount of effort is currently being devoted to this task by drafting committees in this country and abroad. When this has been done it may be possible for regulations and codes of practice to allow for better utilization for fuses generally. Meanwhile there is no reason why engineers should not take advantage of the progress which has been made in fuse practice. The test results described indicate some ways in which this can be done.

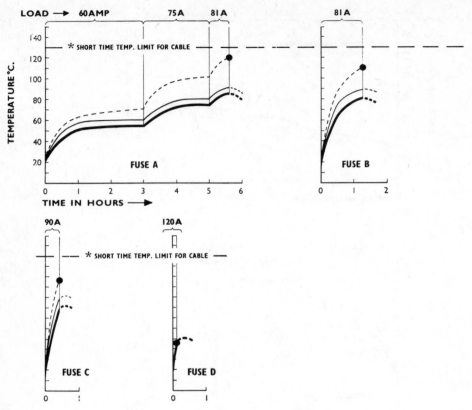

FIG. 8.3
55 A cables protected by 60 A fuse-links. Four 7/.064 in P.V.C. insulated cables in conduit protected by 'English Electric' TIS60 fuse-links. Two fuses in enclosure, fusing factor of single fuse in air approx. 1.5

8.10 Test No. 1

The test represents the case of a large number of small cables bunched in conduit. Ten single-core P.V.C. insulated cables 1/.044 in or (0.0015 in^2) were drawn through about 10 feet of ¾ in steel conduit with two standard bends. The cables were fed from a five-way, single-phase and neutral distribution board with all circuits in series so as to ensure equal loading on all cables both lead and return within the conduit. The cable ends remote from the supply were also weighted to simulate the weight of a 35 ft run of vertically hanging cables (Fig. 8.1).

Temperature measurements of the cable were made continuously at varying distances (1, 3 or 5 ft) from the fuse terminal. Thermocouples for this purpose were inserted under the cable insulation which was then made good and access to the interior of the conduit was suitably devised. The temperature of the fuse contact was also recorded continuously and the results are shown in Fig. 8.2. It should be noted that the nominal fusing factor of the fuses is approximately 1.6, as tested in

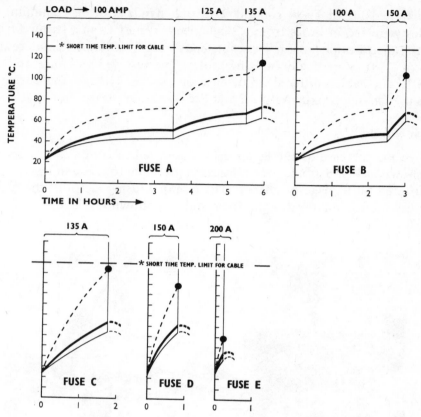

FIG. 8.4
100 A cables protected by 100 A fuse-links. Three 19/.064 in P.V.C.-insulated cables in conduit protected by 'English Electric' TCP100 fuse-links. Three fuses in enclosure, fusing factor of single fuse in air approx. 1.5

accordance with B.S. 88:1952. The difference between this and the actual results obtained is entirely due to the relatively higher temperature of the cable and the effects of the enclosure; the main factors being the heat from the cable tails and the effect of the several fully loaded fuses in close proximity to each other.

It should also be noted that the rating assigned to these cables has been given as 8 A, but this is high compared to the rating in the thirteenth edition of the I.E.E. Wiring Regulations, which assigned a rating of only 5 A. In spite of this the cables suffered no damage or even any detectable change during several tests which were carried out in succession on the same samples.

8.11 Test No. 2

This test represents the case of cables of intermediate rating bunched in conduit. Four single-core 0.0225 in² (7/.064 in.) P.V.C.-insulated cables were run in approximately 10 ft of 1¼ in conduit with two standard bends of minimum radius. The cables were fed in series from a single-phase supply from a two-way single-pole and neutral distribution board. The cable ends were weighted to simulate the weight of a run of 35 ft of vertically hanging cables. The test rig was similar to that shown in Fig. 8.1. The cables suffered no damage or change. The nominal fusing factor of the fuses used was approximately 1.5 and the results are shown in Fig. 8.3.

8.12 Test No. 3

This represents the case of the larger cables bunched in conduit. Test arrangements were as shown in Fig. 8.1, and are identical in principle to those in tests 1 and 2. The cables suffered no damage or change. The nominal fusing factor of the fuses used in this case was approximately 1.5 and the results are shown in Fig. 8.4.

FIG. 8.5
Test rig for bunched P.V.C.-insulated cables in air: A, supply cables; B, fuse box; C, cables under test; D, cable ends; E, weight; F, temperature recorder; G, thermocouples

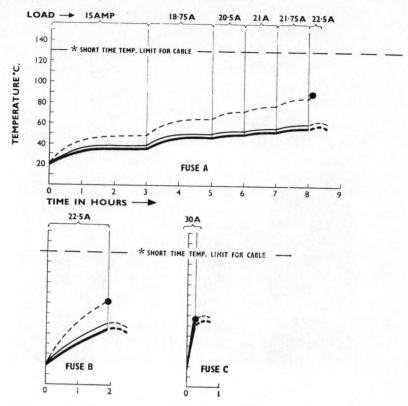

FIG. 8.6

15 A cables protected by 15 A fuse-links. Three twin-core 3/.036 in P.V.C.-insulated cables in free air protected by 'English Electric' NIT15 fuse-links. Three fuses in enclosure, fusing factor of single fuse in air approx. 1.6

8.13 Test No. 4

The test represents the case of bunched twin cables in air. The cables were fed from a standard three-way single-pole and neutral distribution board through standard glands. The cable run passed over two bends of a specified minimum radius and the short-circuited ends were weighted to simulate the weight of a run of 35 ft of vertically hanging cables. Otherwise the conditions were generally as for test 1, and the test rig is shown in Fig. 8.5 and the results in Fig. 8.6. The cables suffered no damage or change and again the nominal fusing factor of the fuse was approximately 1.6.

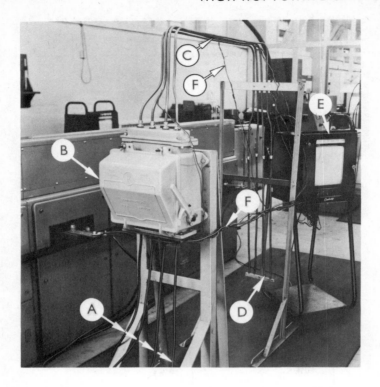

FIG. 8.7
Test rig for mineral-insulated cables: A, supply ends; B, fuse switch;
C, cables under test; D, cable ends; E, temperature recorder; F, thermocouples

8.14 Test No. 5

This test represents the case of mineral insulated cable. A three-phase circuit (3 by 0.04 in^2 and 1 by 0.03 in^2) was fed through a standard 150 A 'Combination' fuse switch. The current was fed from a three-phase supply and the cable ends short-circuited by a star connection. Two bends of minimum radius were included in the cable run which was suitably clamped to an open framework (Fig. 8.7) The cables suffered no damage or change. The nominal fusing factor of the fuse was 1.5 and the results are shown in Fig. 8.8.

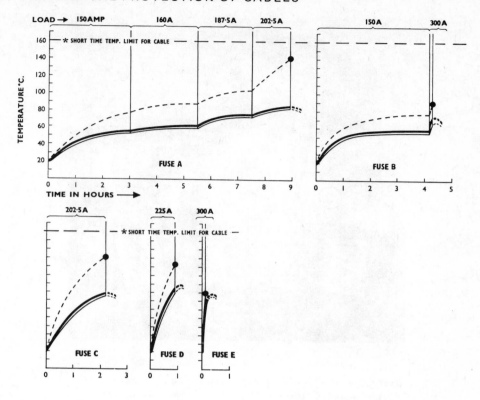

FIG. 8.8

150 A cables protected by 160 A fuse-links. Three 1/.225 in and one 1/.195 in mineral-insulated cables in free air protected by 'English Electric' TF160 fuse-links. Three fuses in enclosure, fusing factor of single fuse in air approx. 1.5.

FIG. 8.9
Test rig for multi-core armoured cables in air: A, supply ends;
B, fuse switch; C, cable under test; D, cable ends; E, weights;
F, temperature recorder; G, thermocouples

8.15 Test No. 6

This represents the case of the larger multi-core cables installed in air. The cable tested was a three-core 0.2 in^2 P.I.L.C.S.W.A. cable connected to a 300 A 'Combination' fuse switch. Fig. 8.9 shows the cable run which passes over a bending mandrel with a diameter of 20 times the cable diameter. A weight corresponding to 35 ft of vertically hanging cable was attached to the cable ends and distributed among the cores in proportion to their cross-sectional areas. Compound was omitted from the cable box, and this was considered to increase severity.

Again, no discernible change could be detected in the cable or cable tails and the results are shown in Fig. 8.10. The nominal fusing factor of the fuses in this case was approximately 1.5.

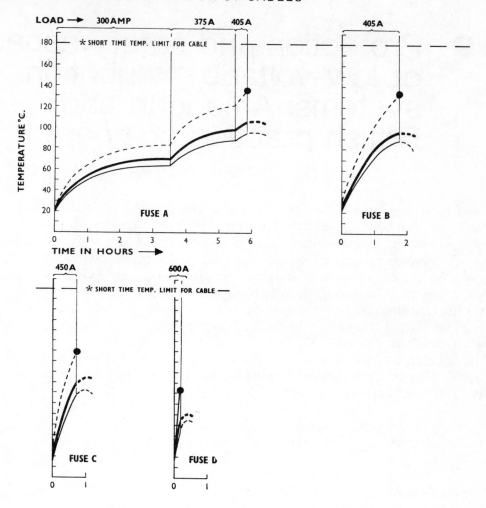

FIG. 8.10
300 A cable protected by 300 A fuse-links. Three core 37/.083 in paper insulated cable in free air protected by 'English Electric' TKF300 fuse-links. Three fuses in enclosure, fusing factor of single fuse in air approx. 1.5

9 Protection and safety aspects of low-voltage distribution systems: American and British practice compared

By and large the American practice in the protection of low-voltage distribution systems has been built around the miniature and moulded case circuit-breaker. For a number of reasons interest in fuses was gaining ground in America. Fault levels had increased sharply, new specifications had emerged and the Underwriters' Laboratories who are the major approvals authority in the United States had begun to concern themselves more actively in fuse practice.

In the changing climate of opinion the organizers of the American Power Conference invited the author to read a paper on aspects of British compared to American practice relevant to the protection of distribution systems.

The benefits derived were mutual and although the respective practices are still divergent, a considerable amount of work has been done to find a common basis and to harmonize ideas and concepts of safety.

Presented to the American Power Conference at Chicago, in April 1965 and originally published by The English Electric Co. Ltd., Fusegear Division in January 1966.

9.1 Introduction

For the purpose of this chapter a low-voltage distribution system includes all the equipment, busbars, switches, circuit-breakers, fuses, cables and terminal devices between the low-voltage terminals of the transformer and those of the load.

The objectives and requirements for the protection of distribution systems are well known and common. Methods by which protection is achieved and the degree of safety demanded may, however, differ according to circumstances, but these differences are usually no more than degrees of emphasis on the same problems. Variations of method occur owing to geographical and climatic conditions, actuarial experience, and the skills available. Different conventions as regards voltage rating and frequency, etc., also arise, but the basic principles are essentially the same. (The British convention is to refer to voltages up to 50 as low voltage and voltages between 50 and 650 as medium voltage.)

In the U.K. as in the U.S.A., it is recognized that for every reason of safety and economy the protective devices in a system should be as intrinsically safe and as reliable as the equipment which they protect, and preferably more so. Also, for the

154

same reasons, installations should be planned to be as free of fault hazard as good planning, adequate materials and mechanical safeguards can make them. The electrical protective devices may then be regarded as insurance against the unavoidable faults which occur due to human error and to other imponderables.

9.2 British statutory requirements and practice

In the case of protective equipment, high performance, reliability, fidelity to characteristics and stability throughout the life of the equipment are prerequisites to good practice. In the U.K. observance of these precepts is required by law as prescribed by the Factories Act. This is administered by government Factory Inspectors who are empowered to inspect all industrial installations at any time of their choosing. Accidents causing injury to personnel due to failure to observe reasonable factors of safety may result in prosecution under common law, apart from any claims for compensation or damages which may ensue.

The quality and performance of electrical equipment is specified by British Standards and there is an increasing tendency at the present time for these Standards to conform with international thinking, mainly through the International Electrotechnical Commission to which the U.S.A. also subscribes.

Installation practice is governed by the Wiring Regulations of the British Institution of Electrical Engineers and other Codes issued by the British Standards Institution. These together are roughly equivalent to the American National Code.

In the U.K. as in the U.S.A., short-circuit fault hazards calling for high interrupting capacity are a major consideration because of the high concentrations of power in the industrial regions. Secondary hazards and problems arising from high fault potentials also exist and not the least of these is the intrinsic reliability and performance of the protective devices employed.

The main antidote in the U.K. to the short-circuit problem has been for very many years the energy-limiting (H.R.C.) fuse which is used almost without exception in every industrial plant and, in some of them, used to the exclusion of other devices. It is not uncommon for industrial distribution systems to be protected solely by energy-limiting (H.R.C) fuses. These are combined with 'on-load' airbreak switches for the control of all low-voltage (or medium-voltage) circuits without recourse to circuit-breakers or other automatic devices. This implies that the H.R.C. fuse can be and is relied upon to provide not only short-circuit protection, but an adequate degree of overload and earth fault protection to ensure the safety of the system. Such schemes have proved entirely satisfactory and over the last thirty years have developed in advance of requirements, notwithstanding the increased complexity of installations and the growth of power sources.

The majority of H.R.C. fuses used in the U.K. have now reached a high degree of sophistication, but this applies to their function rather than to their construction which still remains fundamentally uncomplicated. 'One-time' single characteristic designs are the usual rule. It can be claimed with justification that the high performance, reliability, accuracy and stability obtained from these devices derive directly from their avoidance of complication which in turn simplifies application.

In recent years a good deal of empirical investigation has been made towards the better understanding of the short-circuit withstand of distribution equipment and progress has been made in the design of equipment in this respect. Fault inhibiting properties by means of circuit segregation and by the localization of fault damage have also been put into effect. These improvements, together with the energy-limiting properties of the associated fuse, ensure that unavoidable fault damage is minimal.

The official approval of distribution equipments in the U.K. is provided in ways which are similar in intention to U.S. Underwriters' Laboratories' approval. British users invariably require evidence of certification of rating and performance based on tests by an independent authority. In the short-circuit field these needs are supplied by the Association of Short-Circuit Testing Authorities who control and operate a number of high-power testing stations capable of testing at full power to the highest short-circuit ratings specified, and to a stringent interpretation of the relevant British Standards.

It may be worth noting that public utilities in the U.K. have developed their own methods of distribution and protection independently of the practice in industrial plants, partly because their problems differ and also because they enjoy special privileges in relation to the Factories Act. Nevertheless, they also depend exclusively on H.R.C. fuses for the protection of low-voltage networks.

9.3 Historical: trends of development

The U.K. was already a highly industrial country with centres of large load concentration before the advent of electric power. In the formative years at the beginning of the present century when electricity replaced other forms of power, the electrical load concentrations in the U.K. were probably higher in proportion than in the U.S.A. at that time for purely geographical reasons. This situation led to the early choice of a relatively high standard distribution voltage (415/240 three-phase and neutral) and set the pattern for relatively large 11 kV/415 V transformer units. Thus, the fault levels in industrial plants were commonly of the order of 25 to 30 MVA (30 to 40 kA r.m.s. symmetrical) even during the early 1920s.

H.R.C. fuses were introduced to provide the needed interrupting capacity and in the first place to back up existing circuit-breakers. In the course of time circuit-breakers of the smaller ratings were superseded almost entirely by fuse switches.

The philosophy behind these moves was to provide throughout the plant H.R.C. fuses having interrupting capacities higher than the highest available current likely to be encountered anywhere on the system. Thus, it became unnecessary for the operators of the plant to calculate or even consider the prevailing fault levels. Since all the protective devices were of equal performance supplies could be extended and transformers enlarged or repositioned without affecting the safety of the installation as a whole. Safety was thus ensured irrespective of the state of knowledge of the

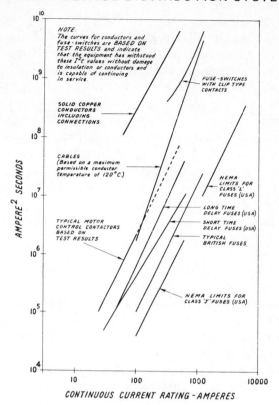

FIG. 9.1
I^2t withstand of equipment (British) and
let-through of fuses

operators who in some cases might be ignorant of the dangers inherent in the situation.

Fuse designs have advanced to still higher values of interrupting capacity and have attained the position in which interrupting capacity as such is no longer a major problem in fuse design, at least for those experienced in the art. This statement does not of course preclude the need for full proof of performance, particularly at the higher values, and there is no substitute for testing at full power under the most onerous conditions which can occur in service.

During the last 10–15 years the emphasis in fuse practice has moved beyond interrupting capacity towards meeting the demand for the minimization of fault damage. This has become more acute as fault levels have risen. The main reasons for the move, however, have been that faults have increased numerically in proportion to the increase of load concentrations and also because of the greater economic dependence of industry upon the service which electrical power supply provides — prolonged outages are expensive.

The most serious faults are arcing faults which by their nature are more damaging to equipment and more dangerous (as regards personnel and fire hazard) than any other type of fault. It is in this field particularly that the energy limiting action of the H.R.C. fuse, together with the sensible design of distribution equipment, has reduced hazards to negligible proportions. An examination of the official accident and fire statistics in the U.K. shows that the accidents which still occur are almost invariably due to human error, interference by outside agencies or inadequate short-circuit rating. Failure of properly approved equipment is rare.

Since the H.R.C. fuse plays so large a part in British distribution practice it is relevant to consider the technical attitudes affecting fuse practice. Many of these attitudes are common to both the U.S.A. and the U.K. although it is appreciated that high-interrupting-capacity fuses are not used in the U.S.A. in the same high proportion as in the U.K.

9.4 Non-deterioration

Great emphasis is laid in the U.K. on the ability of any protective device to resist ageing or any change which might impair its performance. Such ability is essential for sealed devices which cannot be checked or recalibrated after installation and the H.R.C. fuse comes within this category. H.R.C. fuses can be and are designed and rated to resist any form of deterioration due to loading, transient overloads and through short-circuits over a life span of thirty years or more. Full life tests conducted on a particular British design over 20—25 years have proved this to be so and the design parameters leading to this result are well documented.

9.5 Discrimination and selectivity

For economic reasons discrimination is essential and depends both on the accuracy of the fuse and its mode of operation. Discrimination is relatively simple at the lower orders of available current but becomes more critical at the higher available currents above the cut-off threshold.

For discrimination between two fuses in series, the significant factor is the ratio between the melting $I^2 t$ of the larger fuse and the total (melting plus arcing) of the smaller. This factor is a function of the fuse element design, the method of arc extinction and the method of voltage rating. With a manufacturing accuracy of ±5% (i.e. current for a given time) a discrimination ratio within 2:1 between the major and minor fuses is achieved by at least one British design at the highest values of available current, and 1.2:1 at the lower available currents.

9.6 Interrupting capacity and I²t

Interrupting capacity may be considered as relating to the total energy (melting $I^2 t$ plus arcing $I^2 t$) which the fuse can absorb, and it thus bears relation to the physical dimensions of the fuse. Similarly, current rating also relates to dimensions. Since dimensions in the U.K. have become standardized by common consent, it follows that optimum let-through $I^2 t$ in relation to current rating has become established by usage. It is also known, by usage and investigational tests, that the withstand of

British cables, switches and other equipment is well above the let-through of contemporary fuses installed to protect them.

Experience in the U.K. has shown that any cables or equipment which will not withstand the let-through of existing fuses is inadequately designed because the $I^2 t$ let-through is not the limiting factor. Fig. 9.1 shows typical comparisons which illustrate this point. It is considered to be a doubtful proposition to design cables and switches down to minimum withstand values related to the minimum $I^2 t$ let-through attainable with fuses. At these values the mechanical and load carrying capabilities of the equipment must become suspect.

Thus, $I^2 t$ data is freely supplied by British fuse manufacturers for the purpose of judging discrimination and for guidance in respect of those special circumstances where $I^2 t$ let-through is known to be a critical factor. It has not so far been considered necessary to prescribe $I^2 t$ as a limiting factor for the purposes of standardization or official approval of fuses.

Fig. 9.2 shows the extent to which $I^2 t$ values may be varied by design according to the requirements of the application. To adopt the lowest values for distribution systems would be unnecessarily restrictive and is undesirable.

British designs of H.R.C. fuses are manufactured in large quantities to comply with British and Commonwealth Standards and also to N.E.M.A. Standards (up to 200 kA r.m.s. symmetrical) — in many voltages and current ratings.

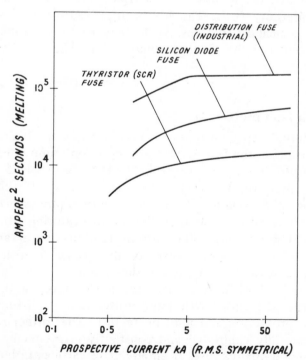

FIG. 9.2
Comparison in $I^2 t$ values of 100 A fuses for various applications

9.7 Voltage rating and frequency

Because of the standard line voltage of 415 V in the U.K., it is usual to rate and test fuses at 440 V, this being the highest system voltage allowable by statute. Although H.R.C. fuses are voltage sensitive it is economically possible for a.c. duty to assign a range of voltages to a fuse which may have been designed for 440 V.

The British mining industry uses a voltage of 550 for underground installations, and it is quite often possible to use fuses of standard dimensions suitably tested and rated for this purpose.

One factor to be considered in this respect is the arc voltage but this does not usually cause difficulty. Firstly, because fuses can be designed to limit overvoltages during arcing to low values and, secondly, because insulation levels on low-voltage systems are relatively high.

It is recognized that the a.c. performance of a fuse bears little relation to its d.c. performance and for this reason separate categories of duty and separate tests are prescribed. High power d.c. testing facilities of the order of 100 kA at 600 V and relatively long time constant have been available in the U.K. for many years and fuses have been regularly tested to this order of performance. It is expected that all fuses claiming d.c. performance should have been tested at full power to be considered safe.

Within wide limits fuses are not sensitive to frequency and the difference between the American 60 cycle and the British 50 cycle frequency is not significant in the performance of contemporary fuses. This has been proved by comprehensive tests which confirm the theoretical calculations. At the higher frequency the peak current is higher by a few percent but the $I^2 t$ let-through is less. The net result in terms of stress is about equal.

9.8 Overload and earth fault protection

In general, the minimum fusing currents of British fuses are higher for a given current rating than is usual in the U.S.A. The ratio between minimum fusing current and rated current is approximately 1.5 to 1.6 in the U.K. against 1.25 to 1.35 in the U.S.A. The result in practice, however, is similar because of the conventions concerning installation and loading. Fuses are temperature sensitive and the temperature in service is usually higher than that attained during rating tests. Thus, the rated (or nominal) minimum fusing current is a maximum and the actual minimum fusing current in service is lower. For this reason, British fuses provide overload protection to cables and other parts of the system.

By the same reasoning, H.R.C. fuses are also relied upon in the majority of industrial plants in the U.K. for earth fault protection. A distinction is made between earth leakage protection and fault protection. The former requires highly sensitive equipment carefully calibrated and constantly maintained. The latter can easily be achieved by good-quality fuses if an adequate earth path back to the star point of the transformer is provided.

CABLE CONNECTIONS   ARC CHUTES

SHROU   SHROUDING OF CONTACTS   ON/OFF INDICATOR

RETRACTABLE HANDLE

DOOR INTERLOCK

FUSES MOUNTED ON SWITCH BLADE ASSEMBLY   INCHES   0 1 2 3 4 5 6 7 8 9 10 11 12

FIG. 9.3
1200 A 500 V 'on load' fuse switch chassis arranged for mounting into equipment ('English Electric')

Controversy in the U.K. concerning the merits of solid earthing or otherwise has continued for many years but the majority of industrial and public utility engineers favour solid earthing on the grounds of simplicity. No compromise is made here with safety because it is proved statistically that the reliability of the fuse/solid earthing concept is greater overall than that of the more complicated alternative devices.

9.9 Fuse switches: disconnects

Fuse switches according to British Standards are used for controlling distributor cables and motor circuits and are required to have making and breaking capabilities appropriate to the duty.

Fuse switches must be tested for making on to a fault of a magnitude at least equal to the breaking capacity rating of the associated fuse. In addition, they should be capable of making and breaking currents equal to several times full load at low-power factor so as to be safe for breaking the stalled rotor current of motors or for cases where they can be closed on to a restricted fault and then opened before the back-up protection has had time to operate.

A popular design in widespread use in the U.K. accommodates the fuses on the moving blade assembly of the switch and this provides safe access to the fuse as well as a double break in the circuit which eliminates danger from feedback from interconnected circuits. Fuse switches of this variety are made in all current ratings from 30 to 1600 A. Fig. 9.3 shows a typical 1200 A fuse switch chassis designed for building into a variety of equipments.

9.10 Switch panels: control centres

The layout of switch panels is important in regard to safety and protection. Short-circuit strengths of busbar structures are frequently specified by the British user. The manufacturer must show proof of compliance by carrying out full scale tests as required. Short-circuit capacity presents no particular problem where fuses

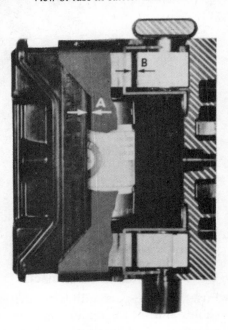

View of fuse in carrier and base.

Base sectioned to show position of shrouding on contacts and busbars. Gap 'A' closes before contacts 'B' engage.

FIG. 9.4
H.R.C. distribution fuse board ('English Electric' Red Spot)

protect the busbar zone and this is a common practice in the U.K. It is recognized that the vulnerable parts of the switchboard equipments are connections, jumpers (between busbars and switches), small wiring, etc. These points are specifically recognized.

Particular attention has been paid to the back-up protection of motor control contactors which are often associated with fuse switches in the same assembly. Load centres are fairly common and many of these incorporate dry-type, air-cooled transformers in the same package.

Another trend which is now incorporated in the best designs is the complete shrouding and insulation of all live parts to prevent accidental contact by personnel. Such measures not only include the shrouding of switch and fuse contacts and other parts accessible when changing fuses, but also include busbars, jumpers, interconnectors and terminals. Mechanical and electrical interlocking is also applied where specified. Fig. 9.4 shows how shrouding has been applied to a distribution fuse board of a type in widespread use in the U.K. This principle is carried through to the largest switch panels.

10 Progress in the development of H.R.C. Fuses

The achievements which marked fuse development in the early 1960s were those relating to fault energy limitation, which is a capability unique to the fuse. There was also a considerable extension of fuse application in the high-voltage field.

The minimization of fault damage resulting from the introduction of H.R.C. fuses and leading to a reduction of 'down-time' in industry brought considerable economic advantages. The most interesting developments, however, in energy limitation, from a technological point of view, were those associated with the protection of semiconductor or 'solid-state' devices. The effective economic use of these devices depends upon the degree to which they can be protected against damaging overcurrents. The problem is critical because they are extremely vulnerable to fault damage. The only form of protection which is rapid and sensitive enough to afford protection, particularly against short-circuits, is the H.R.C. fuse. Special forms of rapid and ultra rapid acting fuses emerged for this purpose and these posed challenging problems to the fuse designer.

In the high-voltage sphere fuses became necessary to an increasing extent to back up circuit-breakers of inadequate breaking or rupturing capacity. In association with simple 'on load' switches they also provided an alternative to the circuit-breaker, particularly for supply networks. The growth in the high-voltage fuse business was not confined to the U.K. and was marked by the introduction of new international standards.

Based on a lecture given by the author to the Birmingham Electric Club in February 1966 and originally published in the *'English Electric' Journal,* May 1966.

10.1 Introduction

During the last few years fuse developments have advanced rapidly in several directions. New duties have continued to arise and there is no sign that the pace is likely to slacken. On the one hand, new applications have emerged, such as those concerned with the protection of semiconductor devices. These have called for new design techniques and have brought a new dimension into fuse technology. On the other hand, the duties required of the more ordinary fuses, such as those already established in industrial and supply networks, have substantially increased. Fuse developments have kept pace with these demands and, in some cases, have anticipated them to the extent that new concepts in system planning and in the design of equipments, have been made possible.

When discussing the progress in this field, it is tempting to dwell on the newer and more dramatic developments but to do this would be to ignore the tremendously important advances in the performance of everyday H.R.C. fuses: advances which are just as exciting technically, and which are in the aggregate of more immediate economic importance.

Advances have been made in the area of short-circuit and other testing. Better understanding of the transient behaviour of systems and circuits has been followed by improved methods of simulating such conditions in the test laboratory, where measuring techniques have kept pace with the requirements which have arisen. Much of this technology is now being reflected in national and international standards, but beyond these there is a great variety of other duties for which fuses have been developed and in which they are now being regularly used.

The widening scope of fuses has demanded a greater variety of data to facilitate their proper use. Much more accurate information on new parameters has become an everyday necessity. In the more sophisticated applications the fuse has to be closely matched with the equipment it protects and must be co-ordinated with other protective devices in the same system. Thus, there is a need for data in respect of both fuse and equipment to be adequate and in related terms.

The duty of a fuse is twofold: to carry varying loads throughout its life, and to interrupt fault currents. Seen from the fuse element designer's point of view, these two roles are mutually incompatible because a large cross-sectional area of conductor or element is desirable for cool running, whereas a small cross section is to be preferred for rapid interruption. These two opposing requirements sum up the dilemma facing the fuse designer. A measure of the success which has been achieved in resolving it is evident in H.R.C. fuses now being marketed, in which it is possible to claim increases in current, voltage and rupturing capacity ratings at the same time.

10.2 Rating

The growth of electricity as a service has led inevitably to the demand for fuses which are bigger and better in every respect.

The largest medium voltage fuse listed in B.S. 88 is of 1,200 A capacity, but for many years now ratings much larger than this have been in regular use in the U.K. There is, in fact, no technical limit to the current ratings which can be met. When the limit of current rating has been reached in one package, there is no reason why fuses should not be accommodated in several packages suitably paralleled (see Fig. 10.2). A good deal of research and careful testing has been done to substantiate this principle and the main factors upon which it depends are well documented. It is, of course, necessary to be certain that the fuses connected in parallel are manufactured to predetermined tolerances and that they are disposed in the circuit to take account of the magnetic forces which may effect current sharing under both normal and fault conditions. Care must also be taken to see that the connections by which the fuses are paralleled are properly effective, and this is usually best done by the manufacturer. Larger fuses in a single package can be made without technical

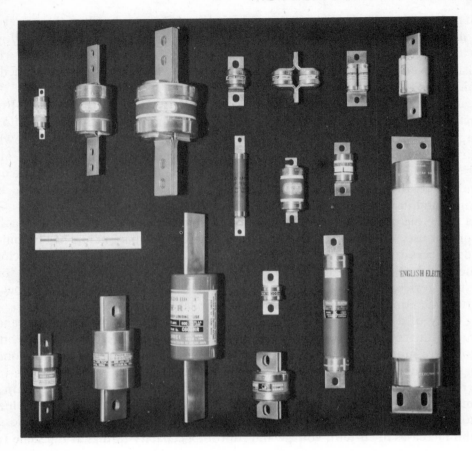

FIG. 10.1
*A representative selection from the many ranges of 'English Electric' H.R.C. fuses
developed for an increasing variety of uses*

difficulty when the demand justifies it. Fuses of several thousand amperes rating are
not uncommon and the main point to be made is that such fuses are both feasible
and useful.

Voltage ratings as with current ratings are limited more by demand and
commercial viability than purely technical considerations. With fuses, voltage rating
is especially important because the fuse is by its nature voltage sensitive. For normal
distribution voltages up to and including 11 kV, H.R.C. fuses have been available for
many years, but it is only during the last two or three years that the demand for the
higher voltage ratings has shown significant increase, so far as the U.K. is concerned.
Coincident with this demand there has been a good deal of research and development which has resulted in new types of fuses having performances far beyond those
which have been previously envisaged at such voltages. Again, higher voltage ratings
than 11 kV are entirely feasible.

FIG. 10.2
*A typical example of a parallel fuse assembly for
the protection of a 3000 hp motor circuit rated at
600 A, 3.3 kV, 250 MVA*

British Standards have always required that H.R.C. fuses should be capable of interrupting fault currents from the lower orders of fault current to the highest short-circuit values, a much more onerous task than that of a fuse designed to interrupt the higher currents only. British Standards also assume that a fuse is capable of withstanding the system recovery voltage indefinitely after it has blown and tests are prescribed to prove this capability. Recent achievements with high-voltage fuses should be judged in the context of these factors.

Because there has been relatively little difficulty in meeting all the requirements of service in medium-voltage a.c. systems, the problems affecting voltage in d.c. systems are not always appreciated. The d.c. rating of fuses is entirely different from the a.c. rating, even though the same fuse may be used for the two jobs.

10.3 Rupturing capacity

Rupturing or breaking capacity is the property from which the H.R.C. fuse derives its name and for which it was originally introduced. There is no doubt that the best

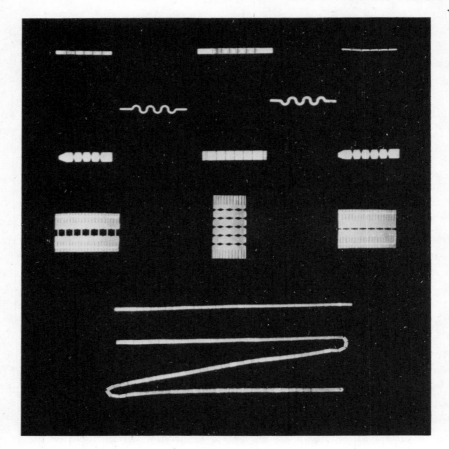

FIG. 10.3
Fuse elements: some of the varied designs which have been developed to meet the conflicting requirements of cool running and higher interrupting capacity

designs owe their continuing success to the fact that they have always been able to provide rupturing capacities beyond the requirements of service and to do so economically. A very brief examination of the performances available will illustrate this point. The present British Standard for fuses lists a maximum category of 46 kA r.m.s. symmetrical (35 MVA) at 440 V, but in a new draft recently distributed to industry for comment, the values go up to 80 kA.

The Canadian Standard issued as long ago as 1952, calls for 100 kA r.m.s. asymmetrical at 600 V and the American N.E.M.A. Standard introduced in October 1959 calls for 200 kA r.m.s. symmetrical, also at 600 V. 'English Electric' H.R.C. fuses to both the Canadian and American Standards were first produced in the U.K. almost as soon as the respective specifications were issued, and were the first in the world to meet these standards. Occasionally, there is argument that the higher values, particularly those called for in the American Standard, are much higher than are required in practice. Such arguments ignore the fact that fuses can be produced economically for these performances, and that in doing so they provide a very adequate margin of safety and open the way towards a new and more economical approach to system planning. It is also a fact that the higher fault levels for which such fuses are required do exist and are increasing.

In the high-voltage field the current British Standard calls for 250 MVA at 11 kV. An 'English Electric' design has been proved to 750 MVA for some years, and it is evident that rupturing capacity as a parameter of performance will not limit even the most exacting applications for some time to come. Moreover, when higher values are required and testing equipment becomes available, appropriate fuse designs can be made available.

10.4 Basis of rating

In all discussions concerning the rating of electrical equipment it is necessary always to consider what rating really means. A 100 A fuse is a fuse which will carry 100 A continuously under prescribed conditions without exceeding a prescribed temperature rise. Since it is obviously impossible to prescribe conditions which correspond to every circumstance in service, it is necessary to make an arbitrary choice. This basis needs to be considered very carefully, particularly at the present time when the severity of service duty is increasing rapidly. The specifications which were satisfactory even ten years ago may no longer be adequate under present conditions. National fuse standards have to be revised periodically on this account and a good deal of new thinking has recently been put into them, both in the U.K. and by the I.E.C. (International Electrotechnical Commission).

Since current rating manifests itself as a thermal condition, it follows that fuse development must proceed in two directions, firstly to reduce the watts loss to prevent excess heating, and secondly to improve the thermal withstand of the materials used in the construction. There is a good deal of evidence in some present designs that progress has been made in both these directions.

Voltage rating is similarly open to scrutiny. The usual method of assigning a voltage rating is to test at a voltage in excess of the rating and to require that recovery voltage equal to the applied voltage should be maintained for a prescribed time after the fuse has interrupted the fault. But there are also transient conditions to be taken into account, not only in relation to the natural phenomena of the circuit but also in relation to the mode of operation of the fuse itself. It cannot be assumed that a fuse rated for a given voltage is suitable for a system having a lower voltage unless it is known that the transient overvoltage produced by the fuse when it interrupts is within the insulation level of the lower-voltage system. This point is particularly relevant in the high-voltage field. Modern design has made it possible so to control the arc voltage of fuses that the same fuse may be suitable for a very wide range of system voltages.

Rupturing capacity rating also requires qualification. It is expressed normally in terms of kiloamps and voltage and, following a convention previously adopted for

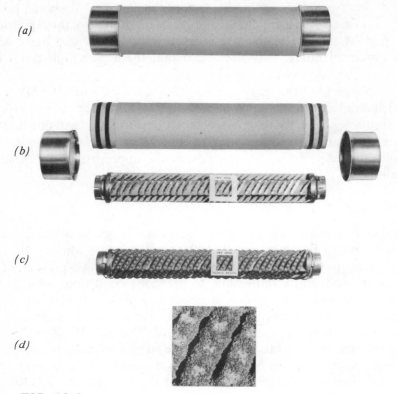

(a)

(b)

(c)

(d)

FIG. 10.4
Views of 11 kV, 750 MVA high-voltage fuse-link, showing:
(a) complete fuse-link; (b) dismantled fuse-link (without filler);
(c) 'blown' element; (d) close-up of 'blown' element to show
regularity of resulting fulgurite formation

three-phase circuit-breakers, is often quoted in terms of MVA. Actually, in specifications, these terms are also qualified by frequency, making or arcing angles, amplitude factor, arcing and inductive energies and control of wave shape generally. The advances which have been made in rupturing capacity rating have been concerned not only with increasing the maximum values but also in evaluating those intermediate values where arc energy stresses are at their most severe.

10.5 Stability

The ability of an H.R.C. fuse to remain stable in performance, and particularly in its time/current characteristic, throughout its life is obviously important from the point of view of protection and safety. This capability is more important in fuses than in some other forms of protective device because the fuse cannot be recalibrated or refurbished in service except by complete replacement, and replacement seldom, if ever, occurs until a fuse has blown. Fuses must, therefore, be able to carry loads and

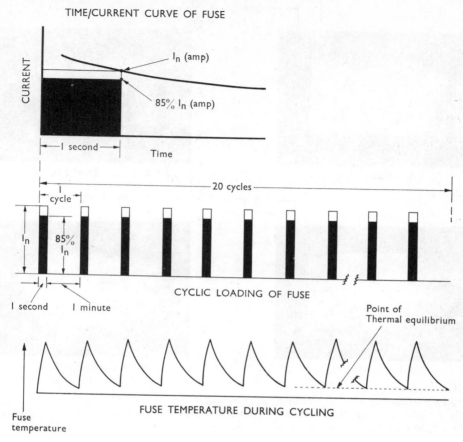

FIG. 10.5
Example of a formalized fuse test to demonstrate the ability of a fuse to withstand repeated motor-starting

transient overloads during a life of many years and then be capable of safely and accurately clearing a short circuit according to the performance claimed for it when new. This property of non-deterioration is essential to stability and has to be positively designed into the fuse.

Considerable progress on this aspect has been made, involving investigations into the behaviour of metals and other materials and into many other physical problems. Prolonged studies have been made of fits and tolerances and other constructional problems involving the fitting together of dissimilar materials, and thermal and heat flow patterns as well as mechanical strengths and chemical compatibilities. Great ingenuity has been employed in devising accelerated ageing tests to simulate the effects of long and arduous life but the most convincing proofs are still to be found in actual service experience. More than ten years ago extensive researches were carried out in connection with fuses which had actually been in service for periods

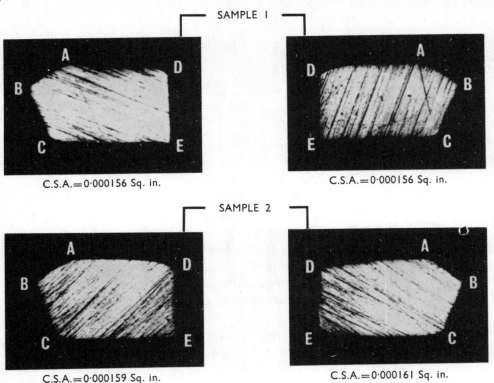

FIG. 10.6
Example of quality control, involving microscopic examination of guillotined silver strip used for fuse elements (micro-sections of the two ends of sample strips to check accuracy of cross-sectional area and consistency of shape

between 18 and 25 years up to that time. The results of these researches, which proved non-deterioration as a fact, were published at the time*. Since then, research on these lines has continued.

Some of the most trying conditions to be found in practice are those affecting fuses in motor or welding circuits where substantial starting surges occur frequently, and where the load pattern is complex and irregular. One of the main problems facing the applications engineer is the evaluation of these conditions in terms which are meaningful to the fuse designer. Once such information is forthcoming then fuses can be prescribed without difficulty to avoid deterioration.

The question of non-deterioration becomes even more important with the growth of highly interconnected and sequence controlled systems. A large generating station or a large automated production process can be jeopardized if the device protecting an essential service is unreliable. Fuses can be made to interrupt circuits safely when faults occur, but it is equally essential that they should be made to stay intact on all occasions other than when a fault is present.

* See chapter 2.

The property of non-deterioration refers to the inherent ability of the fuse to resist deterioration. Correct choice and application is still important, even though fuses can be made to withstand some abuse. It is useful in this context to be able to predict how far a fuse can withstand excessive overloading, and with the best designs this is possible.

10.6 Accuracy

Accuracy in a fuse is required for purposes of discrimination and co-ordination. Accuracy is primarily a function of manufacturing technique and quality control, but it is the initial design which dictates the facility by which a particular degree of accuracy can be achieved economically. The accuracy achievable in the normal H.R.C. fuse is high in relation to that of other forms of overcurrent protection, and the accuracy achievable in fuses made for special purposes is, in some cases, equal to the accuracy of the available instruments used to measure them. This means that the performance of the H.R.C. fuse under given conditions can be accurately predicted, and the fuse can be manufactured to perform faithfully within the limits set by its published performance data.

The fact that a fuse is accurate in itself does not absolve the user from choosing it carefully with regard to the circumstance under which it is expected to work. It is a thermal device which is affected by the thermal state of its environment, and knowledge of environmental effects is a prerequisite to the calculation of its performance. Fortunately, these effects are not serious in ordinary usage and they can often be a safeguard in the case of overload faults. The fuse simply adjusts itself to the thermal conditions and provides closer protection. Where it blows as a result of being too small for the duty, the fuse will 'fail to safety' if it has been properly designed with this end in view.

It is in the protection of semiconductor devices (see Fig. 10.7) that the accuracy of fuses reaches the highest point of present achievement. In these applications the margin between normal operating conditions and those which would cause irreparable damage are extremely fine. The only device which can afford protection accurately enough to work within this margin, and to do so economically, is the fuse. The output obtainable from many semiconductor devices is determined by the measure of protection which the fuse can provide, and the more refined the operation of a fuse can be made to be, the greater are the prospects of using semiconductor devices to economic advantage.

Obviously this work has called for and warranted a large amount of research and development, not only in the fuses themselves, and this has been considerable, but also in the precise evaluation of circuit conditions. The complexity of the mathematics involved makes this a fruitful field for the employment of computer techniques and a good deal of work has been done along these lines.

The knowledge of fuse and circuit behaviour is a preliminary to the ultimate object, which is to manufacture large quantities of fuses which can be relied upon to

FIG. 10.7
*Banks of silicon diodes with protecting H.R.C. fuse-links and auxiliary trip
fuses mounted in a large 'English Electric' 350 V, 24 000 A rectifier
equipment for the British Aluminium Co.*

give protection according to the design predictions. Given an understanding of the
physical nature of all the materials involved and having adopted manufacturing
techniques to produce the results required, consistency of the product then becomes
a function of quality control. This is another field in which new methods and
techniques have been evolved.

10.7 Test resources

Approvals tests at full power and under simulated service conditions are necessary to prove that design and manufacture of fuses are combining to produce the results required. Unremitting inspection and tests are necessary to maintain the standard of performance claimed.

Very considerable resources in test facilities are now required to support fuse technology and production. They must be as great in power output as any system in which fuses are to be used. They must also be capable of producing and measuring any one of the great variety of conditions which may occur

10.8 Fault limitation

Of all the virtues which the H.R.C. fuse possesses, that of limiting fault stresses is the most noteworthy because in this respect the H.R.C. fuse is unique as a protective device. The initial demand for rupturing capacity as such has long been superseded by a demand for a form of protection which not only interrupts the fault but also minimizes the damage arising from it. This has been followed yet again by a demand for protection in which the degree of fault limitation can also be prescribed. In the ordinary industrial distribution system the first considerations are those of preventing danger to personnel and avoiding fire risk. The secondary considerations are those of minimizing damage to the equipment protected, although in practice the latter is closely related to the former.

Fuse protection can be related to the short-circuit withstand of cables, busbars, switches and other control equipment in such a way that the fault energy let-through of the fuse comes well within the short-circuit withstand of the equipment. This implies that both the let-through of the fuse and the withstand of the equipment is known in terms of electromagnetic, thermal and other stresses. In the case of fuses this information is readily available but, unfortunately, this is not true for all the equipment which the fuse protects. The only remedy up to the present time has been to test the combination of fuse and equipment at full power. This is relatively simple, though rather expensive, for 'bolted' faults on such items as cables, switches and busbar structures, but is not so simple where it is necessary to evaluate fault arcing and explosion effects. Work has recently been carried out in this connection, and most instances which can occur in service can be reasonably assessed in terms of information now available.

The advantages of fuse protection in respect of fault energy limitation have not yet been fully exploited and it is in this area, more than any other, that fuse design is forging ahead in advance of the market. The ability to limit current and thus minimize electromagnetic and other mechanical stress, and the limiting of fault energy, so controlling thermal stresses, have been amply demonstrated. Now attention is being turned to the rate at which energy is released and distributed between the fault and the fuse itself. A given amount of energy released slowly creates a thermal stress which may be of no consequence, but the same energy released very rapidly creates an explosion. The dynamics of explosive arcing faults are under active investigation.

At the other end of the scale from the problems of preventing explosions and other excessive stresses, there are also the problems related to the avoidance of stress to a much finer degree, such as in the protection of semiconductor devices. The scale of achievement in this respect can be illustrated by an actual example. An industrial 100 A H.R.C. fuse will limit the fault energy in a typical case to about one-hundredth of what the energy would be if the fuse were not present. The let-through of a fuse designed for silicon diode protection reduces the let-through to about one-quarter of that of the industrial fuse. The fuse designed for thyristor protection in turn limits the let-through to one quarter of that of the diode fuse.

A fuse designed for one specialist application is not necessarily suitable for other applications. Fault energy limitation is only one of the parameters which has to be taken into account in making a suitable choice of fuse, and different applications call for emphasis on different parameters. There are considerable differences for instance between the requirements for industrial systems and public supply networks. Indeed, in the case of the latter, an excess of fault energy limitation is sometimes considered to be a nuisance because it makes the job of finding cable faults more difficult. Fuses in supply networks are not chosen critically in regard to $I^2 t$ because fire risk with underground cables or even overhead lines is not a primary hazard. Continuity of supply is more important, and it is for this reason that larger fuses with greater overload capacity but less sensitivity are usually preferred.

10.9 Time/current characteristics

The prominence which has been given to the short-circuit capabilities of H.R.C. fuses has sometimes obscured their capabilities in other directions. The fact that the majority of fuses in distribution systems has for years been expected to provide overload protection to cables and other conductors has often been ignored. Ever since H.R.C. fuses were first introduced they have been able to provide overload protection to a sufficient degree to make industrial and other systems safe. With the advent of plastic cables, the matter of overload protection has become more critical, but there has been no difficulty in showing quite conclusively that existing H.R.C. fuses can meet the requirements which have arisen. It has been necessary to review British Standards in this respect to achieve more precise interpretation of characteristics such as fusing factor. Work has also been done to rationalize fuse designs to meet the new conditions more effectively but no fundamental changes have been necessary.

10.10 Other factors

Although it is necessary in an objective study to consider the various parameters separately, it is obvious that these parameters have to be combined to produce the

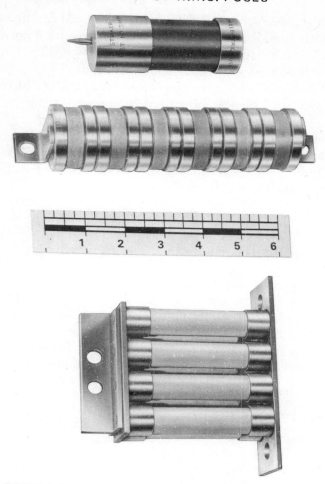

FIG. 10.8
*Typical arrangements of special fuse-links developed for the
protection of rotating rectifiers built into large alternator
rotors*

optimum performance required in a given fuse. Different applications require
different emphasis between parameters and the number of combinations which are
possible are almost infinite. This means that the potentialities for fuse protection are
still very great.

At the present time there are powerful influences towards standardization of
those fuses required for established uses. The flexibility of fuse design can greatly
assist this trend if used in the proper context.

The versatility of modern H.R.C. fuses could not be better illustrated than by the
manner in which they have been applied for the protection of rotating rectifiers used
for alternator excitation. These include fuses which are mounted integrally with
large alternator rotors (see Fig. 10.8). The fuses are thus subjected to the accelera-

tion and other mechanical forces associated with high-speed rotating plant. Fuses are now available which have been tested for correct operation when subjected to accelerating and centrifugal forces of the order of 5000—7000 g. The electrical duty is complex because the transient voltages under conditions of fault are influenced by the mechanical as well as the electrical and magnetic behaviour of the machine. The size and mass of the fuse must be minimized to counteract the centrifugal effects, and forced cooling is usually applied to obtain maximum rating.

The research effort which goes into these sophisticated designs is eventually assimilated into ordinary fuses to benefit the industrial user. This is a continuing process in which the organizations having direct experience of all facets of the fuse industry are able to make the most significant contributions.

11 Fuses: some quality and reliability considerations

1967 was designated as a 'National Quality and Reliability Year', sponsored by the Government through the United Kingdom National Productivity Council. The purpose of this exercise was to maintain and improve the quality of British goods to enhance export prospects. A great number of products were critically reviewed at this time, H.R.C. fuses amongst them.

The quality and reliability of fuses must necessarily be of a high order because of their essential function as a safety device. Safety devices must, in the natural order of things, be intrinsically more reliable than the system which they protect, otherwise they may tend to increase the very hazards they are provided to prevent. Fuses fall into a special category in this respect because they cannot be finally tested for their essential functions without 'blowing' or destroying them. Designs are proved by type tests and thereafter it is encumbent upon the manufacturer to ensure that every fuse he makes is physically identical to the tested prototype and has identical performance.

The measures taken to control quality to the necessary degree to ensure complete fidelity of fuses are therefore particularly important.

Based on Mr Jacks' chairman's address to the Mersey and North Wales Centre of the Institution of Electrical Engineers in October 1966 and originally published in *Electronics and Power*, January 1967.

11.1 Introduction

Questions concerning the quality and reliability of fuses are being actively discussed both nationally and internationally. British Standards are being revised, the International Electrotechnical Commission is drafting recommendations for both high- and low-voltage fuses, and similar committees are hard at work in the European Economic Community, E.F.T.A., North America and every other industrialized country in the world, including the Commonwealth countries.

Inevitably, there are divergences of opinion concerning the philosophies surrounding fuse protection and the way in which these philosophies have manifested themselves in specifications and practice. The high-rupturing-capacity (H.R.C.) fuse was pioneered in the U.K., and having built up a vast store of experience, the U.K. has continued to play a leading part in the branch of electrical engineering which concerns fuses. However, the U.K. has no monopoly of ideas affecting fuse protection, and these are now pouring out on a global scale.

The 'Quality and reliability year' being conducted by the U.K. National Productivity Council provides an incentive for reviewing the present position in terms by which British ideas of fuses and fuse protection may be valued.

11.2 Functional product

It is first necessary to establish the meanings of quality and reliability in the context of fuse protection, because obviously these words have different meanings for different purposes. For a functional product such as a fuse, quality means complete suitability for the purpose for which the fuse is designed. Reliability is the maintenance of essential quality throughout the life of the fuse in service. These virtues are not so obvious in electrical products, because of the abstruse nature of electricity. Nothing exemplifies this dilemma more than the H.R.C. fuse, which is simple in conception as a piece of hardware, but is very complex in function.

This chapter is concerned principally with the question of how quality and reliability are determined and proved, rather than how they are achieved in manufacture. In practice, these questions are, of course, indivisible, and one is no more or less important than the other. Fuses are not any more or less difficult to manufacture than any other quantity-produced article, but, because their duty is to provide safety, they must be made to standards of quality consistent with their responsibilities in this role.

The quality of any product is related to both its function and economy. The balance between these two must be reasonably maintained, and different products obviously require different orders of quality. From an economic point of view, an excess of quality in a given situation is as undesirable as an insufficiency: although it is perhaps fair to comment that, where the function is one of providing safety, an insufficiency often causes greater difficulty to the user in the long term.

11.3 Service requirements

Economy does not imply the cheapening of a product; it means that one is working more closely to the limits of the specified or known requirements, rather than achieving more performance than is necessary. Economy also means that the service requirements are clearly recognized so as to achieve a proper balance between the various duties which may be involved over the whole field of usage. The same criteria also apply to reliability, as it applies to the functional life of a product. Where the service-life expectancy can be specified, as in many durable consumer products, the requisite order of reliability can be designed and built into the product.

What standards of quality and reliability apply to fuses? The deciding factors are those which recognize first and foremost that fuses are protective devices. A purely economic approach to the question may lead to difficulties, because the economics of safety involve human factors and are not easy to define. Similarly, an actuarial approach is of doubtful value, because the situation is constantly advancing beyond the bounds of past experience. The economics of fuse design must anticipate future trends as well as assimilating past experience.

The economics of fuses as a means of protecting plant and property, as distinct from protecting personnel, are easier to see. Fuses and fusegear are equipment which form part of a service; and continuity of the service is almost always of more economic importance than marginal savings on the service equipment. An electrical-distribution system as a whole usually represents a small capital cost, compared to the cost of the capital plant or establishment which it serves. This is why the most stringent standards of manufacturing and testing are justified for protective equipment and for fuses in particular.

11.4 Unavoidable faults

The order of quality appropriate to fuses has also to be seen from an engineering point of view, relative to the rest of the equipment with which they are associated in the system. The proper function of fuses, or any other protective devices added to a system, should be to deal with those unavoidable faults which may occur in spite of the best system planning and installation. Prevention is better than cure, and the first line of defence against danger in a system is the inherent quality of the system itself. Fuses must be at least as high in quality and reliability as the system they protect; it would be folly to weaken an otherwise good installation because of the inadequacy of the protective devices installed. Conversely, it is poor practice to depend on protective devices to bolster up a skimped installation, except as a compelling expediency.

Fuses, unlike other protective devices, are not expected to require (and do not receive) maintenance, recalibration or refurbishing during their service lives. It is therefore all the more important that they should be able to remain as effective up to the end of their useful lives as they were at the beginning, bearing in mind that their lifespans are completely unpredictable. They may be hours or decades, depending fortuitously on when a fault may occur.

If the criterion of quality is that of complete suitability for a purpose, this begs the question: what is the purpose? Or, in other words: what are the requirements which have to be met? These are the questions which exercise the minds of fuse designers and those people who are concerned with the writing of specifications and standards.

It might seem that all the secrets of the electric circuit are already known; indeed, there is no lack of theoretical knowledge concerning its behaviour. But the circumstances which arise in practice are infinitely variable, and there is still a large and fruitful field for investigation, to reconcile theory with practice. A good deal of work is still to be done on the analysis of the behaviour of circuits and systems, particularly under the disturbed conditions which apply when short circuits occur.

Because the requirements affecting fuse protection are so infinitely varied, it is necessary to bracket the main parameters into practical combinations to cover the majority of normal applications. This is done in most fuse standards, so that fuses can be applied, in many cases by rule of thumb, by users who may have only a sparse knowledge of the electrical principles underlying the performance of the fuses they use. This accentuates the need for careful definition of the conventional parameters used and, in particular, for a very careful choice of the basis on which fuses are rated. The conventional parameters – current, voltage, frequency and rupturing or breaking capacity – may still mean different things according to different specifications.

Other essential parameters relating to electromagnetic or thermal stresses in the fuse or the system also require careful definition. It has to be remembered that these phenomena cannot be defined purely in terms of circuit requirements, because, when the fuse is interrupting a short circuit, the transient behaviour of the system may be modified by the mode of operation of the fuse itself. Many problems which arise from this behaviour defy theoretical solution and have to be solved by painstaking empirical investigation.

In addition to the more involved problems concerned with the active function – when the fuse is interrupting a circuit – there are also those relating to the passive function – when the fuse is doing no more than carrying normal load currents.

11.5 Environmental conditions

Environmental conditions, for instance, can affect fuse performance in several ways, and must therefore be adequately specified. Ambient temperature and atmospheric

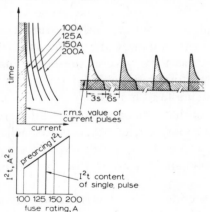

FIG. 11.1
Evaluation of irregular loads in relation to choice of suitable fuse. The r.m.s. value of the pulses appears to require a 125 A fuse; $I^2 t$ of a single pulse is greater than the pre-arcing $I^2 t$ of the 125 A fuse; therefore, the choice is a 150 A fuse

conditions are the principal effects to be considered in most locations, and these can vary widely. A fuse which cannot cope within practical limits of tolerance of temperature, altitude, humidity and atmospheric contamination is of little use. It must be designed to resist any change in its physical condition which could in any way impair its electrical performance.

Factors such as chemical compatibility of associated materials, or the lack of it under ordinary environmental conditions, may be significant in this respect. The effect of thermal cycling on dissimilar metals, and other materials which may be involved in fuse construction, can also give rise to interesting questions. These issues can govern the range of duties over which a given fuse may be designed to apply.

When a fuse is carrying normal load currents, it is just a conductor, and, although by design, it is a 'weak link', it must not jeopardize the circuit by getting hot or otherwise becoming unstable. A 100 A fuse is one which will carry 100 A continuously under specified conditions without exceeding a specified temperature rise. This qualification is important, because any difference between specified conditions and those which apply in service may call for an adjustment in the rating. Usually the range of service applicability is widened by providing a contingency within the specification, or by the designer providing extra capabilities to satisfy particular requirements.

Load currents are not always steady and regular. The problem in choosing a fuse for an intermittent or irregular load is usually the difficulty of being able to estimate the load accurately, rather than any difficulty in evaluating the performance of the fuse (Fig. 11.1) A fuse can be supplied to a predetermined characteristic, and has a 'memory', or temperature time constant, which is easily defined.

11.6 Active functions

A fuse must be capable of safely interrupting overcurrents under any of the conditions which come within its rating. Rupturing or breaking capacity is normally

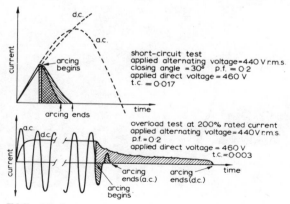

FIG. 11.2
Comparison of arcing stresses in a.c. and d.c. circuits for a typical 100 A industrial fuse

FIG. 11.3
One of four similar machine halls in a short-circuit testing station, an indication of the size of plant required for modern fuse testing

defined in terms of prospective current and voltage, and power factor for a.c. or time constant for d.c. As short circuits are random occurrences, a prospective alternating current is subject to various degrees of asymmetry, depending on the point on the voltage wave at which the short circuit is initiated. The point on the voltage wave when arcing commences is significant. The severity of the fault, in terms of arc energy related to both the fuse and the circuit during the arcing period, is influenced by the instantaneous system voltage during arcing. Fuses can be voltage-sensitive when rated very closely to the limits of their inherent breaking capacity.

For a.c. fuses, the most severe stress does not necessarily occur at the highest prospective currents. The critical current corresponding to the maximum arc energy released in a fuse, which is the condition of maximum thermal stress, can be estimated for each particular design. For d.c. fuses, the most critical and severe duty may be that which occurs on the lower orders of overcurrent. For this and other reasons, it is always better to segregate the a.c. performance of a fuse completely from the d.c. performance, and to consider them as completely separate categories of duty (Fig. 11.2).

In common with other interrupting devices, fuses are liable to produce an overvoltage during the arcing period. It is essential that this voltage, normally called the arc voltage, should keep within the insulation level of the system to which the fuses are applied. In certain cases, the arc voltage may also have to be considered in relation to the characteristics of other voltage-sensitive devices, such as surge diverters, which may be installed in the same system.

11.7 Safety margins

When the fuse has interrupted the circuit, it must be capable of withstanding the recovery voltage in the system, and may be required to remain in service for an indefinite period, during which it must not restrike or allow possibly troublesome leakage currents. In a few milliseconds immediately after the circuit has been interrupted, the rate of rise of recovery voltage and its amplitude are irregular. The degree to which this phenomenon has to be taken into account depends largely on the safety margins allowed in voltage rating.

Any purchaser requires assurance that a product is suitable for the purpose for which he buys it, and, in a functional product, it is desirable to show empirical proof of performance. Where fuses are concerned, this is imperative, because, as has been shown, it is difficult to define the requirements except in functional terms. It would obviously be uneconomical and unreasonable to try to test for every circumstance which could or might arise.

The problem resolves itself into that of deciding on those combinations of circumstances which are most representative of the severest service conditions. Techniques have been devised to simulate service conditions as faithfully as possible and as adequately as necessary. The degree to which tests can be standardized, so that they can be repeated in different locations at different times on a basis of strict comparability, is also important.

Loading tests, to verify the current rating of fuses, are more laborious and time-consuming than technically difficult; but, because they seem simple, they can be misleading unless properly conducted. The criterion is temperature rise; thermal interactions with adjacent equipment make it impossible to test a fuse in isolation, and thus all such tests are, in basic conception, comparative tests. The temperature time constants of large fuses may be several hours, and the task of maintaining a steady load current and stable thermal conditions over such an extended period is not an easy one. Loading tests are sometimes required to simulate intermittent loads, e.g. the load curve of a supply network, or the pulsing type of load associated with welding or automatic inching of motors. In other cases, it may be necessary to reproduce particular and irregular waveshapes or recurring transients.

Short-circuit tests are expensive, because they have to be carried out at full power. The power required to test, for instance, a 750 MVA 11 kV fuse requires generators and transformers of a size which would not be out of place in a moderate-sized power station (Fig. 11.3). Even medium-voltage fuses require power of the same order to produce the high prospective currents now specified. One standard requires 200 kA r.m.s. symmetrical at 600 V, with an asymmetry factor which makes the peak asymmetrical value more than 500 kA.

11.8 Control and measurement

Power capacity of the test plant, expensive as it is, is not the main consideration. Control and measurement are often more important in producing test results convincing enough to ensure the adequacy of the fuses tested. Control is necessary to produce the conditions which simulate the desired degree of severity, and accurate and co-ordinated measurement is needed for the integrity of the results.

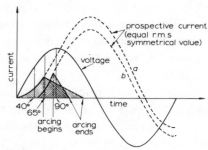

FIG. 11.4
*Tests for rated breaking capacity,
showing effect of arcing angle: curve a,
arcing to commence between 40° and 65°
on voltage wave (maximum thermal
stress); curve b, arcing to commence
between 65° and 90° on voltage wave
(maximum electromagnetic stress)*

Nothing exemplifies the degree of control necessary in short-circuit testing more than the requirements concerning point-on-wave selection, which, in common with all the other parameters and quantities involved, requires to be controlled within fine limits. A fuse may interrupt a short circuit in no more than a few milliseconds; but during this time the sequence of operations has to be timed and controlled to satisfy the specifications for test at maximum severity.

Some tests are needed in which the arcing of the fuse must commence between specified limits of arcing angle to produce the highest electromagnetic stresses on the one hand and the highest thermal stresses on the other (Fig. 11.4). This requires an accurate knowledge of the fuse characteristics and careful control of the making angle. If control of this order were not possible, this would mean a very large increase in the number of test shots required; or it might give results of doubtful validity.

11.9 Loading devices

Testing at the lower orders of current (about 200% of current rating) may seem at first to be less of a task than testing at the very high currents of hundreds of times current rating. This is not necessarily so. At the very high currents, the time of operation is short; at the lower currents, the time is long, and, although the power source can be smaller, there is the problem of providing loading devices large enough to dissipate the energy. A 100 A 11 kV fuse which takes about 1 h to blow at 180 A would require a loading of about 2 MW, which would have to be dissipated for 1 h. Loading problems of this magnitude are formidable, and have led to methods of synthetic testing, in which the fuse is made to carry the appropriate current at low voltage until just before it blows; the full service voltage is then applied. The control needed to ensure valid tests is again of a high order.

The techniques needed to control large machines under short-circuit conditions and the methods of adjusting circuit constants and maintaining current and voltage have been developed to a highly sophisticated degree. The scale of testing operations is continually increasing; with this increase come further problems affecting control, which has to be no less critical, in spite of the vast scale of the power to be handled.

As with control, the measurement of large currents (of perhaps hundreds of kiloamperes) at service voltages is not simple. The value of certification-of-approval tests depends entirely on the integrity of measurement; and almost every quantity involved requires special facilities to ensure fidelity of results. Even then, a good deal still depends on the skill and experience of the test engineers, who, in the last resort, often have to use their professional discretion in deciding the validity or otherwise of a particular result.

11.10 Range of applicability

A combination of tests at maximum rating, critical and overload currents, specified power factor and point-on-wave may be reasonably expected to ensure that a fuse

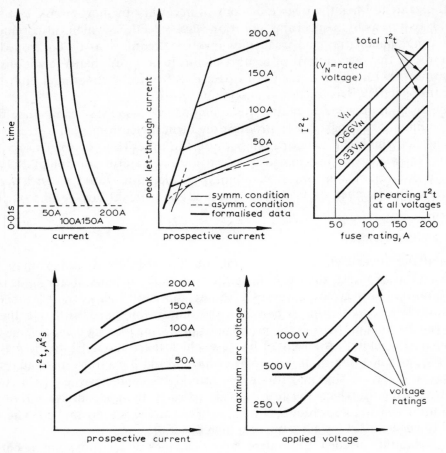

FIG. 11.5
Typical fuse-application data

will deal satisfactorily with all magnitudes of fault on the system, with a given voltage and frequency, corresponding to those for which a fuse is rated. But these tests in themselves do not ensure satisfactory operation on other voltages and frequencies. Fortunately, the possibilities of extending the range of duties can be fairly clearly stated.

For instance, an 11 kV fuse will be suitable for a 6.6 kV system if it has been specifically designed, as some fuses are, to produce an arc voltage which does not exceed that which a 6.6 kV system will withstand.

A fuse suitable for a 50 Hz system will usually work correctly on a 60 Hz system and vice versa, provided that it is designed with this end in view. Also, a 50 Hz fuse can give satisfactory performance at frequencies higher than 60 Hz. Indeed, fuses designed for 50 Hz have been tested and proved for actual service where the frequency can reach 2 kHz at prospective currents of several hundred kiloamperes. It is an interesting fact that the thermal response of an H.R.C. fuse is such that it will still exhibit cut-off and interrupt the circuit safely even at these very high rates of rise of prospective current.

11.11 Characteristics

The data required for the application of fuses are extending in range and detail because of the increasing variety of the applications themselves (Fig. 11.5). It follows that fuses must be made accurately to conform to published characteristics, which in turn must be comprehensive enough for practical use. Work in this sphere of fuse technology has been greatly facilitated by modern methods of computation; many examples could be quoted of how computers have been put to good use in this direction.

11.12 Statutory implications

The premium which has been put on safety, as implied in existing legislation and for economic reasons, calls for the highest standards of fuse performance and application. In the U.K., the interpretation of the Factories Acts and contractual obligations as regards safety implied in the Merchandising Acts are symptomatic of this need. British Standards, the I.E.E. Wiring Regulations, and codes of practice in Britain and elsewhere are constantly being revised to meet changing practices and conditions. In other countries, technical standards for fuses are mandatory by statute; and, because they have to be capable of legal interpretation, the demand for more explicit standards is urgent. Even so, official standards will never be the last word; they should, and usually do, follow the best practice, and should be constantly revised to keep pace with it.

12 Fuse performance data for modern applications

Fuses, in common with other protective devices, are for the most part applied in service by people who do not have, or do not require to have, deep knowledge of the technological factors involved. The means by which fuses are identified, i.e., conventional ratings and published characteristics etc., are necessarily reduced to simple terms to avoid unnecessary complications in day-to-day applications. Safety margins are applied to cover contingencies and there is a need to keep these under constant review in the light of changing circumstances.

The simplifying of performance data for the benefit of electricians in the field places greater responsibility on specifying authorities concerned with the presentation and interpretation of data. Practising engineers need to know the limits to which standard data may be applied so as to be able to judge safety limits and exercise discretion in those cases where standard conditions may not apply.

Based on a paper presented at the E.R.A. Distribution Conference, Edinburgh, October 1967 and originally published by The English Electric Co. Ltd., Fusegear Division in December 1967.

12.1 Introduction

The reliability of fuses derives to a large extent from the simplicity of their construction and also from the manner in which their ratings and performance data* have been rationalized to permit uncomplicated usage by unskilled persons. But this rationalization is not a casual matter. It is based, or should be based, upon the most comprehensive knowledge of fuse technology and practice. Current, voltage and breaking capacity ratings together with fusing factor and nominal time/current characteristics are usually sufficient for most normal applications, but implicit in them is an enormous amount of information which is not always apparent to the non-expert. Fuses are actually much more versatile than is generally recognized in normal usage and more than the existing standards and official specifications suggest. They are constantly increasing in variety and application far beyond and well in advance of published standards. Thus, the demand for more explicit performance data increases proportionally — and is being met. The more specialized data is of course of interest only to those who need to apply it and does not concern the majority of users except as an academic interest.

The success of fuses and the standards of safety which have been achieved by them up to the present time have depended very largely upon the safety margins inherent

* The word 'data' is used throughout as a collective singular noun.

in existing specifications. It is necessary now to avoid a situation in which this success may breed complacency which may inhibit the more technically-advanced applications of fuses and conceal their value in relation to new problems now arising. The purpose of this chapter is to draw attention to some of these new requirements and to review existing practices and conventions in the light of present-day knowledge.

Although circuit constants and parameters obey natural laws and do not change fundamentally, the relationships between them are constantly changing in significance and emphasis. At the same time, fuses, in common with other protective devices, are being applied more critically to achieve closer protection. Co-ordination between fuses themselves and with other protective devices is becoming more critical. It follows naturally that there is now a greater demand for fuse data to a greater degree of accuracy and in greater detail than heretofore. Consistent with this demand is the need for more meaningful information in like terms concerning the fault and overload withstand of equipment which fuses are expected to protect.

The presentation of data is particularly important. The practical comparisons between published characteristics of fuses and other devices and also with the behaviour of equipment to be protected is only possible if such information is available in directly comparable terms. Obviously such comparisons are greatly facilitated if the information is presented on a common basis. This is a simple fact which is more easily stated than achieved. Even in its most comprehensive form the presentation of data must of necessity be a rationalization to some degree. It is impossible to represent every circumstance which may arise. The correct interpretation for any given circumstance depends to a large extent upon the manner in which data is presented, bearing in mind the state of knowledge of persons who have to apply it. The skill required in selecting a proper method of presentation may be greater in those cases where data is to be interpreted by non-expert users.

12.2 Basis of rating

The behaviour of a fuse in a given situation depends in the first place on the method by which it is rated. In common with other protective devices, a fuse-link has to be rated for its passive function of carrying current without overheating and also for its active function of interrupting overcurrents. It must also have a voltage rating and comply with limits imposed concerning other circuit parameters which affect severity of duty. These primarily govern the performance of the active function.

The time/current relationship which governs the interruption of a fuse under given circumstances is a phenomenon natural to the fuses, whereas the current rating is an arbitrary value. Fusing factor, which is equally arbitrary, governs the desired margin between load-carrying capability and the minimum current at which the fuse operates on overcurrent. These relatively simple terms have in the past become so conventionalized and familiar that many ordinary users have not appreciated that current rating and fusing factor are not absolute terms or that minimum fusing current and, indeed, time/current characteristics have to be referred to on a conventionalized basis.

It is obviously an advantage if the conventions adopted are common to all designs, and it is for this reason that the recently issued British Standard 88:Part 1:1967 has gone to some lengths to standardize the basis of rating for those fuse-links which come within its scope. This means that in future all fuse characteristics and ratings which purport to comply with this British Standard can be known to be directly comparable.

The basis of current rating is described in B.S. 88:1967 as being that current which a fuse-link will carry continuously without exceeding a specified temperature rise when connected into a specified and dimensioned test rig connected in turn to specified and dimensioned conductors. The ambient and other conditions under which tests have to be carried out to prove current ratings are also specified, having been chosen to represent service conditions to a reasonable degree. But the main point is that where the test conditions may differ from service conditions it is possible to envisage the application of correction factors because the basis of rating itself is known.

The previous issue of B.S. 88 described current rating as that current which the fuse would carry without exceeding specified limits of temperature when fitted into the equipment in which it would normally be used in service. The latest I.E.C. recommendations and other national standards follow this method, which is satisfactory so long as the terms of it are recognized. It means that a fuse rated in one type of equipment cannot be assumed to have the same rating when used in another type of equipment. Existing published data does not always make this point clear, although any misconceptions which may arise are mitigated to some extent where watts-loss limits and other safeguards are imposed. It cannot be claimed however that this method permits free interchangeability between fuses procured from different sources, an objective which is to be desired in the interests of safety.

As with current rating, the basis of breaking-capacity rating is governed in the new British Standard by the use of detailed test rigs. These are designed to produce a circuit configuration which simulates the most onerous conditions of service. In breaking-capacity tests the thermal effects of the test rig are not a criterion. The object of the test rigs is to ensure complete standardization of testing procedures as between testing stations. There is also a further and increasingly important consideration. It is now becoming necessary and quite usual to require data to be published concerning the current- and energy-limiting properties of the fuse in terms of 'let-through' values. These are as dependent upon circuit and environmental conditions as upon the mode of operation of the fuse itself. The transient behaviour of circuits during arc interruption is not simple and, even where it is theoretically understood, the avoidance of unknown variables by adequate specification is essential. It follows therefore that the test conditions upon which published data of this kind depends should be on a standardized basis.

12.3 New relationships

The primary function of the fuse is to interrupt fault currents quickly and safely, but its main virtue over other interruption devices is its ability to limit the current and energy which flow into the fault. It is the exploitation of this capability more than any other aspect of its duties that has required the recognition of new values and new relationships between the fundamental parameters involved.

The degree of current and energy limitation provided by H.R.C. fuses has always been of high order. It is, roughly speaking, a function of the size of the fuse. Conversely, the size of the fuse is determined by the degree to which a fuse can limit the energy which is released in itself. Size and energy limitation are therefore mutually dependent, and fuses which are physically small in relation to the prospective fault current for which they are rated must of necessity give a high order of energy limitation, otherwise they could not function as interrupting devices at all.

Because the degree of limitation has been so high there has been no necessity to apply it very critically. Where, for instance, the choice is between an H.R.C fuse and an alternative device, and where the fuse is capable of limiting fault energy to one or two per cent of the energy which the alternative device would admit, the tolerance on the fuse let-through can be quite wide without affecting the issue. The degree of current and energy limitation may become relatively more important when questions of co-ordination have to be considered. More recently, the application of fuses to the protection of solid-state devices has called for critical degrees of limitation. This has called for a new crop of characteristic curves relating to the various values involved in order to apply fuses to the degree of accuracy required, thus the factors which affect the degree of limitation and the methods by which they can be measured must be duly studied (see Fig. 12.1).

In those circumstances where peak current and $\int i^2 t$ can be used uncritically, the safest convention is to determine the values which apply to the most onerous conditions of service and to ignore the fact that the fuse will admit lesser values under all other conditions (Fig. 12.2). Since the margin between the fuse let-through and the withstand of the equipment which it protects is in these cases extremely wide, the use of maximum values of peak current and $\int i^2 t$ presents no particular problem. In actual service, the maximum values are seldom if ever realized; thus, the use of data in this manner presents a considerable safety margin over and above that which is already apparent from the known margins of protection.

In those cases where fuses must be used more critically in order to provide a closer margin of protection under short-circuit conditions, the convention of using the maximum values may be neither practical nor economical. For these applications it is necessary to know how the let-through of the fuse varies with variations in circuit conditions.

The variation of pre-arcing $\int i^2 t$ with time can be calculated or, at least for a given design, it can be estimated with sufficient accuracy for practical purposes from a knowledge of the fuse element design.

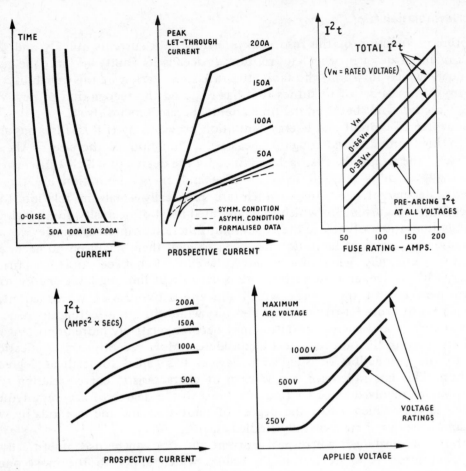

FIG. 12.1
Typical fuse data showing new relationships which have emerged due to the application of fuses for the protection of semiconductor devices

The variation in arcing $\int i^2 t$ is much more complex because it involves not only variation of the circuit constants and the natural and transient behaviour of the circuit, but also the mode of operation of the fuse element and of the method of arc control which is employed. Different fuses operate in different ways and, although the fuse is fairly simple in conception as a piece of hardware, the variations in design are very numerous. Thus, the only reliable way of obtaining $\int i^2 t$ data for practical applications is to do so empirically by testing each design under a sufficient variety of simulated service conditions. Data is required showing the relationship between $\int i^2 t$ (usually rationalized to $I^2 t$) and combinations of applied voltage and prospective current, having regard to the variety of other circuit constants, the incidence of fault initiation and any environmental factors which could significantly affect the results.

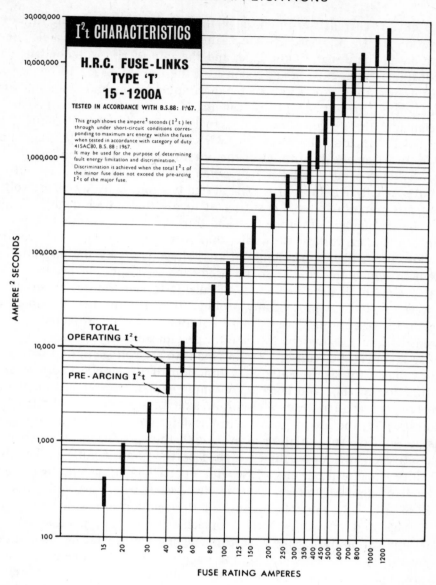

FIG. 12.2
Typical I^2t characteristics.

For purposes of co-ordination it is necessary to differentiate between pre-arcing and arcing values and to obtain a favourable ratio between them; the optimum values are those which give the best ratio for co-ordination while achieving a minimum total because this is the more significant factor in minimizing fault stresses.

Peak or cut-off current is a function of the pre-arcing I^2t and the rate of rise of current produced by the circuit. Thus, the energy required to melt the fuse is dependent upon the design of the fuse, and the rate of rise is dependent upon the circuit constants. It is expedient to consider the pre-arcing I^2t to be constant for very short times; thus, for a known band of applications and assuming sinusoidal conditions, peak current values are calculable. This exercise is of value to the academician and the designer, but for actual use it is usual to compile data from empirical results. Incidentally, pre-arcing I^2t can never be a constant and must vary even down to extremely short times of the order of a few microseconds.

The arc voltage produced by a fuse when it interrupts has to be controlled by design and is usually a compromise. The design techniques by which arc voltages are controlled are numerous and it is sufficient to say that a high degree of control is possible. The arc energy within the fuse and thus the duty upon the fuse is in inverse proportion to the arc voltage allowable; that is to say, the energy is lower if the arc voltage allowable is higher. The limit of arc voltage is set by the insulation level of the system in which the fuse is to be used, and as insulation levels may vary in different systems fuses have to be made available to suit the various situations. Arc voltage does not, in fact, present any problem to either the fuse designer or to the user except in exceptional cases where equipment to be protected is extremely voltage-sensitive, such as may be the case for certain solid-state devices.

The fuse is unlike a circuit-breaker in regard to arc voltage in that with a fuse the arc voltage may vary significantly as a function of the applied voltage as well as being a function of time and current. A simple explanation of this phenomenon is that, whereas the arc length of a circuit-breaker is controlled by the mechanics of the circuit-breaker, the arc length of a fuse is determined by the amount of element which is consumed. The variation in arc length thus produced then depends, for a given inductive energy, upon the incidence of the applied voltage. This is a very useful facility because it permits a fuse with a particular voltage rating to be used on a wide band of applied voltages even in those cases where arc voltage is a critical factor. It is especially valuable because of the relationship which exists between applied voltage and I^2t. One method of achieving minimum values of I^2t for a particular application is to use a fuse of higher voltage rating than the applied voltage of the system. This would not be possible if the higher voltage rating involved a higher arc voltage than the system could tolerate. The desired relationship between arc voltage and applied voltage can be varied in degree according to the design of the fuse. It does not of course follow that all fuse designs have such facility.

The term 'applied voltage' itself covers a multitude of possibilities. A rationalized expression for applied voltage is by no means sufficient to indicate the performance of any interrupting device. What are more important are the instantaneous values of applied voltage and the incidence of these with other physical manifestations during arc interruption. It is normal to rate fuses at voltages corresponding to normal power frequencies or for d.c. (It should be said at once that performance in d.c. circuits

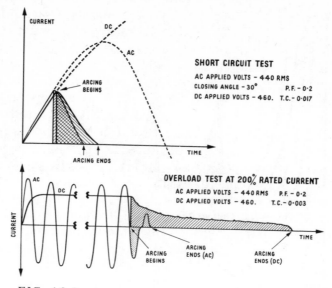

FIG. 12.3

Comparison of arcing stresses experienced by fuse (and equipment protected), showing differences between a.c. and d.c. at different overcurrents

must be regarded as being something entirely separate from performance in a.c. circuits. It is better to consider them as entirely separate entities.) (See Fig. 12.3.) It is then necessary to consider how far an applied a.c. voltage varies from the sinusoidal and how far the d.c. voltage varies from a constant value. Then, in between a.c. and d.c., is a large variety of other voltage conditions involving quite complex wave shapes, some of which may alternate about zero and others which may be unidirectional but of complicated wave shape and discontinuous in character.

Power frequency a.c. tests are done under reasonably sinusoidal conditions and d.c. tests under reasonably steady-state conditions in order to make them universally applicable. The problem then is to translate the conventional data from such tests into terms which are of practical use for unconventional situations. This is an area in which it may be dangerous to generalize and where it it necessary to consider each application on its own merits. A good deal of experience and information has been accumulated to this purpose and methods for representing it in a generalized form must emerge in due course. Meanwhile there is enormous scope for further research, first to analyse circuit behaviour and to then relate the results to fuse performance. The conditions which may arise in circuits with semiconducting characteristics require particular attention along these lines.

12.4 Basic parameters

The basic parameters affecting fuse technology are fairly well known and are to be found in most of the modern fuse standards. But the terms and definitions by which

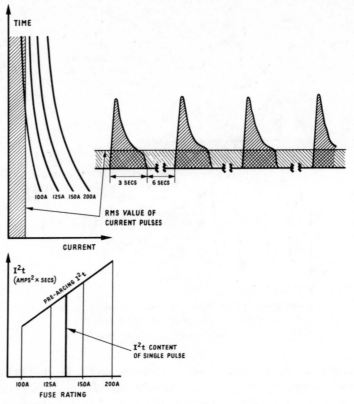

FIG. 12.4

Evaluation of irregular loads for choice of fuse; r.m.s.
value of pulses appears to require 125 A fuse; I^2t of
single pulse is greater than pre-arcing I^2t of 125 A
fuse; therefore, choice is 150 A fuse

the parameters are circumscribed are sufficient only for the practical application in a subjective sense of the specification to which they apply. They may be sufficiently explicit when considered in this context but require a rather closer examination where they are required to be understood in relation to the more advanced applications.

The important criterion in considering the current-carrying capability of the fuse is temperature, and temperature is a main factor upon which current rating is determined. It is simple to specify a limiting temperature which corresponds to the steady state and stabilized condition which is obtained after a uniform current has been flowing for a sufficiently long time. In practice, this seldom if ever occurs. Load currents vary in magnitude and duration and fuses vary as regards temperature time constant. Moreover, the fuse experiences mechanical and other stresses in varying degrees due to temperature variations. One of the problems in practice is to equate the effects of variable loadings with those due to steady-state conditions so as

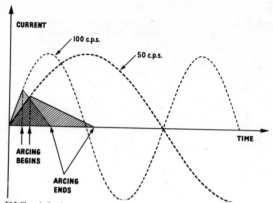

FIG. 12.5

Diagram showing effect of supply frequency on total
$\int i^2 \, dt$ *and cut-off current and total let-through*
(example produces values as described when i^2 *is*
integrated with respect to t)

to judge whether a particular fuse will be proof against deterioration in service.
While it is relatively simple to establish the short-time rating or the no-damage limit
of a fuse corresponding to the shorter times, it is not always so simple to know the
service loads in like terms unless a directly measured record of the load is available.
Even so, a simple r.m.s. evaluation of load current with respect to time does not
necessarily take into account the consequential stresses resulting from transient
peaks of short duration (see Fig. 12.4).

Voltage is a vital parameter because the fuse is a voltage-sensitive device, as indeed
are most circuit interrupters. Voltage rating is chosen as that at which a fuse will
interrupt safely all currents within its breaking-capacity rating within specified limits
of transient behaviour of the circuit and with a margin for contingency. The
minimum voltage at which a given fuse may be applied in a given system may be
proscribed by the arc voltage which the system will stand. The maximum voltage at
which the same fuse may be applied may also be decided by the maximum fault
energy let-through which the system will permit; thus, the voltage rating chosen may
be well within the inherent voltage capability of the fuse where no such limits are
imposed. Arc voltage as such may be varied according to the method of arc control
employed and is a function of design. These points show there is considerable scope
for varying fuse performance as required in relation to the voltage parameters.

Within quite wide limits current-limiting H.R.C. fuses are not affected by
frequency, except in so far as it may modify their let-through characteristics. For
instance, the rate of rise of current at 60 Hz is greater than at 50 Hz. Therefore,
cut-off current of a fuse interrupting the same prospective current under otherwise
identical circuit conditions will be 5 per cent or 6 per cent higher at 60 Hz than at
50 Hz. But because of the incidence of the applied voltage the $I^2 t$ let-through would
be 7 per cent to 8 per cent less. For practical purposes a fuse tested and rated at
50 Hz is satisfactory for 60 Hz duty and vice versa, as is shown both by calculation
and test (see Fig. 12.5).

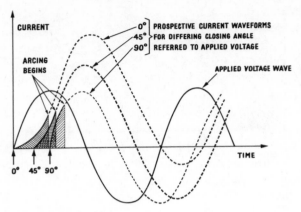

FIG. 12.6
*Effect of closing angle on fuse cut-off current at a
particular prospective current in a fault circuit having
a low power factor*

As the frequency tends towards d.c., the interrupting capability of the fuse at the
lower and medium overcurrents may be less, because it is in these zones that d.c.
duty is the more onerous. A fuse which has been tested and rated at a given
frequency will almost invariably safely interrupt short-circuit faults of higher
frequency because the $I^2 t$ let-through will be reduced, and the fact that the cut-off
current may be higher will not of itself create additional stress upon the fuse. The
thermal response of H.R.C. fuses is such that they will exhibit cut-off at extremely
high frequencies where the prospective current is proportionally high. Tests on
high-voltage fuses have shown quite satisfactory interruption at 2 kHz to 3 kHz with
prospective currents of the order of 500 kA peak and cut-off in the region of 60 kA
to 80 kA.

Contemporary standards and specifications allow tolerances on frequency
between 45 Hz and 62 Hz to cover the majority of power frequency applications.
This is entirely reasonable for current-limiting types of fuse. It does not however
apply to expulsion and other non-current-limiting types.

The breaking-capacity rating of a fuse implies that the fuse should be capable of
breaking all values of current from minimum fusing current to the breaking-capacity
rating under specified conditions of circuit severity. It is well known that the
maximum stresses in the fuse itself do not necessarily occur at the highest values of
prospective current. Generally speaking, for a.c. the maximum severity may coincide
with the condition of maximum arc energy stress which occurs usually within the
zone of currents just beyond those at which the fuse is beginning to exhibit cut-off;
for d.c. the critical zone of currents may well be those just above the minimum
fusing current; for fuses which use organic material in their construction a critical
zone may be experienced at the point where the organic material is subjected to the
highest temperatures, and this may occur within a band of currents between two and
ten times full-load rating.

Medium-voltage and many high-voltage current-limiting fuses designed to British Standards are capable of breaking all currents up to their breaking-capacity rating. Some designs elsewhere in the world cannot. I.E.C. recommendations and also other national standards cater for fuses which have no overload interrupting capability. These are the so-called 'back-up' fuses and are intended for use in conjunction with low-breaking-capacity devices, the duty of which is to provide overload protection. Thus, there are two quite distinct classifications of breaking-capacity rating which need to be clearly understood. The simple use of an MVA or kA value does not in itself convey to the user sufficient information for fuses to be applied safely.

The property of minimizing electromagnetic and mechanical stresses in equipment, resulting from the ability of the fuse to limit fault current, is a variable which depends upon a number of factors. The cut-off current itself depends upon the size of the fuse in relation to the prospective current for which it is rated, the rate of rise of prospective current and the mode of operation of the fuse itself (see Fig,. 12.6). Some fuses are designed to provide a higher degree of current limitation than others.

The definition of peak or cut-off current requires careful interpretation. It does not follow that the peak current during fuse operation coincides with the commencement of arcing, although for many designs of fuse this is the case.

Where a fuse is applied at a voltage very close to its maximum inherent voltage capability, the current may continue to rise during the arcing period to a value considerably in excess of that which is present at the commencement of arcing. Since the criterion is the limitation of electromechanical stresses, and since this is proportional to the square of peak current, this should be understood as meaning the maximum value of current attained during any part of the interrupting period. (Calculation of peak current from pre-arcing $I^2 t$ data is therefore of value only when it is known that the peak value coincides with the commencement of arcing.) The rate of change of current is also very significant in respect of the dynamic forces which may occur and must be duly regarded.

By a similar token, $I^2 t$ also requires scrutiny. In the first place $I^2 t$ is a rationalization of the quantity $\int i^2 dt$. It is a useful comparative term but has no direct entity as a physical manifestation. Several attempts have been made to find a descriptive name for it. The most accurate one, which has been favoured in British circles, is 'specific energy'. An equally accurate though rather cumbersome term is 'joules per ohm'. Yet another term which seems to be finding favour in I.E.C. circles is 'joule integral'. This is a technical misnomer but may find acceptance as a conventional term. Most practising engineers concerned with fuse technology prefer to use the term $I^2 t$ or 'ampere^2 seconds' and to accept that it is a rationalization which is proportional to the energy let-through by a fuse. In American specifications the term $I^2 t$ is regularly used and is described as 'let-thru'.

$I^2 t$ let-through, by whatever name it is called, is an extremely useful concept for comparing the relative abilities of various fuses and other interrupting devices. It does not, however, give any direct indication of the amount of thermal stress which may occur in equipment which a fuse protects. This is proportional to $I^2 R t$, and the

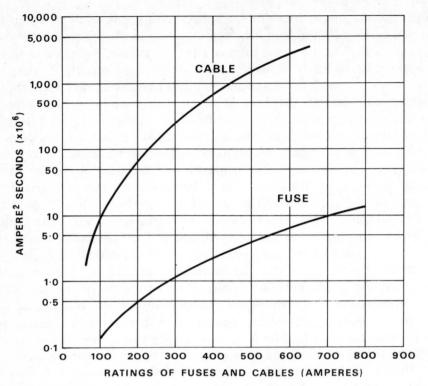

FIG. 12.7
Typical relationship between fuse let-through and cable (no-damage) withstand

stress obviously depends upon the distribution of R in the circuit. Even so, individual pieces of equipment which comprise the circuit may be tested for their ability to withstand a given current for a given time, and this withstand may be expressed in terms of $I^2 t$ in order to relate it to the degree of protection which may be required (see Fig. 12.7).

In the case of arcing faults, other considerations apply because the resistance of the fault is a widely varying factor. On lower-voltage applications the resistance of the fault can be high compared with the circuit impedance, so influencing the fault current and affecting the operation of the fuse. The dynamic effects resulting from the sudden release of thermal energy due to changes in resistance also have to be taken into account. These may give rise to dynamic stresses or even explosion hazards, and $I^2 t$ as a term can give no indication of these phenomena.

The resistance of the fuse (unblown) is very small compared to the resistance of the circuit in which it is normally placed. It has therefore very little significance in relation to the circuit but is a useful value with which to exercise quality control between fuses of like design. There are however a few common misconceptions concerning it. Because it is small it is relatively difficult to measure; in some cases the resistance of the connections to a fuse can be of the same order as the resistance of the fuse-link. It is therefore necessary to define very carefully what is meant by resistance and how it shall be measured. Some specifications require that fuse resistance should be measured in the 'hot' condition, which means that the fuse should have carried its rated current for sufficient time for its temperature to stabilize before the measurement is made. This involves a long and expensive procedure and even so is subject to many inaccuracies unless the current is kept at an extremely steady value during the measuring process. Some specifications have attempted to compromise by allowing measurements to be made within a closely specified short period after the current has begun to flow. This method is fraught with difficulties not only of stabilizing the current but also of taking into account the variety of thermal characteristics in the various designs of fuse. Since the resistance value has no particular significance in the circuit and is required for comparison purposes, the cold resistance measured by applying a negligible small current, and thus avoiding thermal complications, is undoubtedly the most reliable method.

A further term which is coming into regular use is that of 'watts loss'. This is required to ensure that when fuses are installed together in an enclosure the total temperature allowed for the complete equipment is not exceeded. There is logic in this requirement provided that the watts loss of all the other components within the same enclosure are known to the same degree of accuracy, otherwise the logic is doubtful. The problems of measurement are not unlike those relating to the measurement of resistance. What appears to be a straightforward case of measuring simple and fundamental quantities is in fact a rather difficult process. Unless it can be reduced to a fairly simple convention it is reasonable to specify watts loss as being the product of the cold resistance value and the nominal current rating multiplied by a factor which is chosen to take account of the temperature coefficient of resistance of the particular design of fuse under construction.

12.5 Fuse application and co-ordination

Fuse applications are so numerous that it is impossible to do more than indicate trends by drawing upon one or two examples of topical interest.

A very large proportion of all fuses in use are in motor circuits. The fuse provides short-circuit protection and is built into and co-ordinated with motor control gear which provides overload protection. The problem has always been to fix the takeover point between the two forms of protection; the fuse must not blow during motor-starting operations and the contactor must not be called upon to interrupt currents above its breaking capacity. In the past, the choice of fuse and contactor has presented no problem because of the wide margins available. Now the incentives to achieve more efficient utilization of material to save space and reduce cost have tended to reduce these margins significantly. To meet these demands, fuses have been designed to withstand higher horsepower values for a given fuse rating and contactors are available in smaller sizes than previously. At the same time, and particularly in the higher-voltage field, fuses are being used increasingly for the specific purpose of reducing explosion damage due to arcing faults.

Since most of these requirements are mutually incompatible it is necessary to examine the basis upon which the comparisons can be made and to indicate clearly the degree of accuracy which may be expected in relation to each item of equipment used. There are instances where the takeover point between fuse and contactor can be considerably higher than is apparent from a direct comparison of the published curves available. It has to be recognized that the published curves for contactors refer to a normal operating condition. This means that they must be capable of operating without trouble up to their specified breaking capacities. The fuse curve on the other hand denotes a once-and-for-all interrupting condition. There are also other mitigating factors under conditions in which increased loading is applied. The environmental conditions invariably cause the fuse to speed up its operation when called upon to interrupt. Even if the takeover point is chosen with an excess of optimism, the hazard is slight because it will only become apparent if there occurs a fault which coincides precisely with the narrow band of current values which may be in doubt. The statistical probability of such a fault occurring is, of course, very slight and all these factors taken together support the contention that the new pressures towards the more economical use of equipment can be justified — always provided that fuses and contactors can be made to more accurate limits and the data by which their performances are represented are equally accurate.

The greater emphasis on the minimization of fault damage and reduction of let-through I^2t has also led to changes in fuse design in relation to motor circuit duty. Considerable progress has been made along the lines of reducing pre-arcing I^2t while at the same time maintaining, or even improving, the ability of the fuse to withstand legitimate transient overloads. A reduction of the pre-arcing I^2t will, in these cases, of course, also reduce the arcing I^2t as a result of the reduction of inductive energy. Some care must be taken when assessing horsepower withstand to take into account the time factor. The heating effect of a number of successive motor-starting operations when integrated with respect to the relatively long time involved may well come within the time/current characteristic of the chosen fuse, but it is possible that the heating effect of a single starting operation can exceed the pre-arcing I^2t of the fuse. This situation can arise particularly where the starting current of the motor is unusually high and of a peaky nature.

A very similar misconception can arise if the motor starting characteristics (which show the variation of current with respect to true time) are mistakenly super-imposed upon fuse time/current characteristics. To render the motor characteristic in terms which are meaningful to the fuse characteristic, it must be integrated with respect to time and reduced to a single value of r.m.s. current.

In associating fuses with motor control equipment perhaps the most critical problem which arises is that of protecting the thermal elements which are used for out-of-balance or single-phasing protection. The 'no-damage' characteristics or short-time ratings of these follow a fairly normal inverse-time characteristic. It is necessary to ensure that this characteristic is presented on the same basis as that of the fuse which is chosen to protect it. There is no actual difficulty in providing protection, but the choice of a suitable fuse requires some care and a recognition of the differences in approach which may exist between the respective forms of published data.

Problems affecting discrimination between fuses themselves have been fairly well documented. In distribution circuits the best policy is to plan the layout to achieve the maximum ratio between fuses in series, and in practice this presents little difficulty. A rule-of-thumb ratio of 2:1 between successive fuses, which has been popular with electricians for many years, is easily achievable in the majority of cases, but this includes very wide safety margins and does not represent the most favourable ratio which can be achieved. The rule for discrimination upon which this ratio is based — that the total $I^2 t$ of a minor fuse must be less than the pre-arcing $I^2 t$ of the major fuse — is convenient, but is a use of the term $I^2 t$ which is more conventional than scientific. The fuse element responds to energy which is $I^2 R t$ and not merely to $I^2 t$. R is a variable quantity which, in the case of the minor fuse (which blows), varies from its initial value to infinity whereas, in the major fuse, it varies only from its initial value to that which corresponds to a temperature well below melting point. Thus, the energy absorbed by the minor fuse is very much greater than that absorbed by the major fuse. Since the major fuse is usually physically larger than the minor fuse its final temperature can be expected to keep well within that at which deterioration may commence, even when a ratio of less than 2:1 exists.

By far the most interesting and complex problems arising in fuse application and co-ordination are those relating to the use of fuses for the protection of solid-state devices. The fuses used for this purpose are designed on the same principles and obey the same laws as fuses for other purposes. But because they have to be closely matched to the semiconductor devices which they protect, and because these are in great variety and are used under a great variety of circuit conditions, every aspect of the performance of fuses for this purpose has to be intensified, developed and refined (see Fig. 12.8). Such fuses are required to be faster and smaller, with better arc control to keep arc voltage within limits. They must be capable in many cases of withstanding higher loading temperatures and they must perform to greater accuracy. The output and effective utilization of solid-state equipment is very often determined by the degree of protection which can be provided, and in most cases the fuse is the only device which can provide protection at all. This application, more than any other, has provided the incentive for rapid progress in fuse development and the results already achieved in this direction are already having beneficial effects when applied to the more general and traditional usages.

12.6 Fault withstand of equipment

The choice of fuses for protecting a particular piece of equipment presupposes that the characteristics of the fuse can be related to the characteristics of the equipment

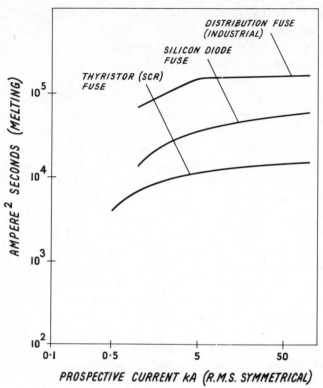

FIG. 12.8
*Comparison in $\int i^2 t$ values of typical 100 A
fuses, showing achievements in development
of fuse design*

in such a way that the fuse will operate before the equipment suffers damage. Where
the margin of protection is a fine one, as is increasingly the case in modern systems,
there is little point in concentrating on the accuracy of the fuse data unless the fault
withstand performance of the equipment to be protected is known in like terms. In
some cases where protection is known to be critical a good deal of trouble is taken
to ascertain the withstand performance but in other cases the information for this
purpose is very sparse.

There is a tendency to uprate equipment such as busbars and cables in terms of
short-circuit performance when current-limiting fuses are used, and this is entirely
reasonable if done with proper regard to all the conditions which can arise. The
maximum thermal stresses in equipment usually coincide with the higher levels of
short-circuit current at which the fuse exhibits cut-off but not necessarily so in every
case. For those equipments in which the heat dissipation is naturally retarded, the
temperature build-up may be the more serious at the lower fault currents. There has
also been a tendency in some quarters to design equipment down to the let-through
values associated with fast-acting fuses and to assume, on the basis of short-circuit

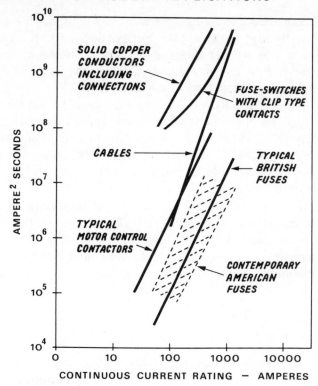

FIG. 12.9

Comparison between fuse let-through and 'no-damage' withstand of various types of equipment, illustrating substantial margins which can exist at the higher currents

tests, that such equipment is safe at all currents. Whether this is true depends on a proper relationship between the known withstand of the equipment and the fuse characteristics.

Generally speaking, if cables, busbars, switches, circuit-breakers and other distribution equipment are rated adequately for load-carrying capability there is usually very little difficulty in providing complete protection by current-limiting fuses (see Fig. 12.9). The margins that have been traditionally accepted for many years provide a considerable margin of safety even down to quite low overloads. It is only when equipment is to be used at very much higher current densities that the problem may become critical. Within this category of equipment is included, in particular, high-density contacts and connections as well as most semiconductor devices. The problems of expressing the withstand of these in meaningful terms is not easy, but it is an increasingly important issue. Fuse designs have advanced to meet these conditions and a good deal is possible. Perhaps the largest problem of all is to educate the intending user into appreciating what the possibilities are.

As would be expected, the problem of protecting against electromagnetic stresses bears little obvious relationship to that of protecting against thermal stresses.

Electromagnetic stresses become more significant at the higher levels of short-circuit current which, when limited by fuses, are proportional to the square of the fuse cut-off current. Since the cut-off current bears some relationship to the size or current rating of the fuse it follows that the electromagnetic stresses are only of importance when the rating of the fuse tends to be large.

The values at which electromagnetic stress predominates over thermal stress often depends upon the type of equipment being considered. For instance, certain types of butt contact may be very adequate thermally but may be subject to contact bounce due to electromagnetic stresses at moderate currents. Many diodes or thyristors can be more susceptible to peak current than to I^2dt because of their semiconducting characteristics.

It is reasonable to claim that the performance and behaviour of fuses can be laid down much more accurately and in greater detail than that of the equipment which fuses protect. There are many areas in which the use of fuses is inhibited because of the lack of information concerning the equipment, and for the same reason it is probable that, occasionally, fuses are wrongly applied in ignorance of the true facts.

12.7 Presentation

The presentation of fuse data is a means of conveying to the user a picture of fuse performance and must be chosen in the context of the application for which the data is required as well as the state of knowledge of the user himself. The manner of presentation not only provides essential facts but also conveys degrees of emphasis and, what is more important, eleminates information which is not necessary for the immediate purpose. The responsibility for providing information in a correct context is considerable. Equally considerable is the responsibility of the user to recognize the context in which the information is presented to him.

Great efforts have been made in recent years to keep fuse data, and the methods by which it is presented, completely up to date with contemporary requirements. The trend is for more data and greater detail. A few general observations and examples may serve to illustrate the philosophies which govern these efforts.

Far-reaching attempts have been made in recent years to standardize the graph paper upon which time/current characteristics are presented. The idea of having a common graph paper for various types of protective devices, including fuses, is attractive. It is not however completely logical for all cases. If, for instance, a fuse curve embracing thousands of seconds and hundreds of kiloamperes is drawn to the same scale as that of another device which is concerned only with a few seconds and a few amperes, the comparison can become ludicrous. It can fail completely to convey the required information to the user.

On the other hand, the idea of having common graph paper for all fuses is much more practicable and this in fact has been virtually achieved both in British Standards and in forthcoming I.E.C. recommendations.

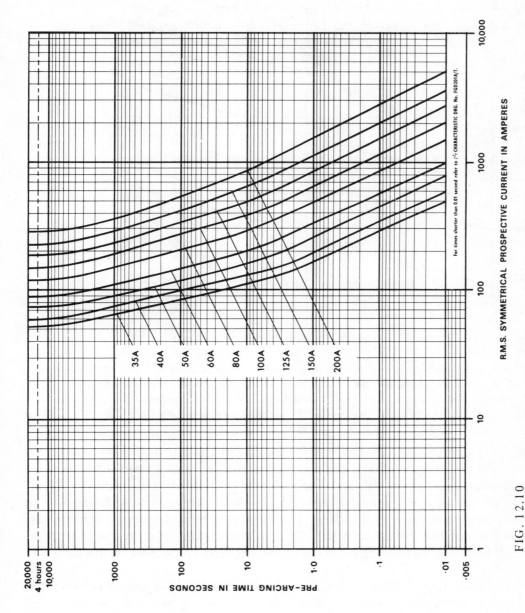

FIG. 12.10

Typical fuse time/current presentation as specified in B.S. 88: 1967.
H.R.C. fuse-links type 'T', 35–200 A. Tolerance on currents shown is ±5%

R.M.S. SYMMETRICAL PROSPECTIVE CURRENT IN AMPERES

PRE-ARCING TIME IN SECONDS

For times shorter than 0.01 second refer to i² CHARACTERISTIC DRG. No. FGD201A/T.

35A
40A
50A
60A
80A
100A
125A
150A
200A

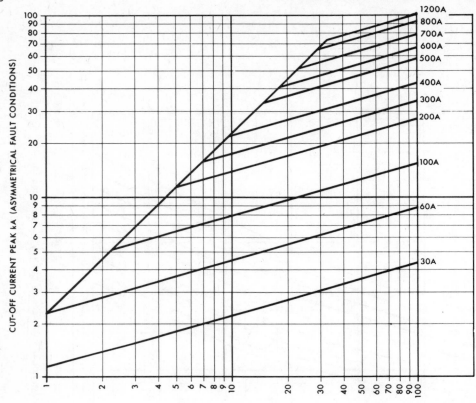

FIG. 12.11
*H.R.C. fuse-links type 'T', 30–1200 A. Typical method of representing fuse cut-off characteristics,
illustrating comparison between r.m.s. symmetrical values of prospective current and peak
asymmetrical values of cut-off current. This graph shows the cut-off currents obtained under
short-circuit conditions when the fuse-links were tested in accordance with B.S. 88: 1967 at 415 V*

The graph paper chosen is a standard International size, both scales are
logarithmic, and the decades of the logarithmic scales are to a fixed dimension and
proportion chosen to give the best pictorial relationship between current and time.
Logarithmic scales are an admirable device for compressing a good deal of
information into a small space. They provide the correct degree of emphasis and
accuracy where it is needed to suit fuse characteristics, but they are one stage
removed from the simpler linear picture and must be recognized in this light (see
Fig. 12.10).

Fuse time/current characteristics are represented either by a single line subject to
a tolerance or by an envelope. Both are satisfactory if they are free from
misconceptions concerning the tolerance limits. It is possible to declare manufacturing tolerances which can be applied in relation to the current values, these being
the independent variables. It is impracticable to apply in this way fixed tolerances to

the time values, which are the dependent variables, because at the lower orders of current a small difference in the value of the current can account for large differences in the time; the converse applies at the higher orders of current. To users familiar with the conventions which have been traditionally adopted for magnetic relays this point is not always obvious. With relays, the time/current relationship is such that the time settings are admissible and discrimination is often defined in terms of time setting only.

Other conventions which tend to confuse the non-expert are those in which fuse behaviour under asymmetrical conditions is referred to the ratings which, by convention, are expressed in symmetrical terms. The presentation of cut-off characteristics is a favourite example in this context.

For a given fuse, assuming a constant value of pre-arcing I^2t, cut-off current is a variable which depends on the magnitude and wave shape of the prospective current, but it is presented as relating to the breaking capacity expressed in r.m.s. symmetrical terms. It requires some knowledge of the conventions involved to fully appreciate what the data really represents. Conventions apart, the values presented are usually taken from actual short-circuit tests involving specified conditions of asymmetry. What is also important is that the user should know the worst that can happen, and the published curves should be set out to represent the information in this light (see Fig. 12.11).

The variation in current values due to different conditions of asymmetry affect also the time/current curve for the times below about one second. For a given value of r.m.s. symmetrical prospective current there is considerable spread in the time values due to the random switching which can occur in normal service, particularly where the circuit is inductive. These variations do not imply that the fuse itself is variable. The fuse is consistent and responds to the conditions imposed upon it (see Fig. 12.12).

Environmental effects become more pronounced with most designs of fuse at the longer times. The time/current characteristic may be affected by variations in ambient temperature, but only in those regions of the curve corresponding to times of several minutes or longer (see Fig. 12.13).

Arising out of the presentation of fuse data are several conventions, the validity of which requires re-examination from time to time. The examples of the 2:1 ratio between successive fuses, used in some quarters to ensure discrimination, is typical. The point has already been made that a degree of discrimination better than this ratio indicates can usually be obtained, but there are other conventions which have been applied to this principle which are incorrect in the converse sense. The idea that the ratio between pre-arcing I^2t and arcing I^2t can be constant is of course erroneous, and this fact needs to be recognized when fuses are used for the protection of critical equipment. It would be convenient if fuse data could be confined to the presentation of pre-arcing I^2t, which would allow the application of a fixed multiplier to obtain the value of total I^2t. Unfortunately, circuit conditions vary so widely that this is not reasonable and may lead to misunderstanding if it is

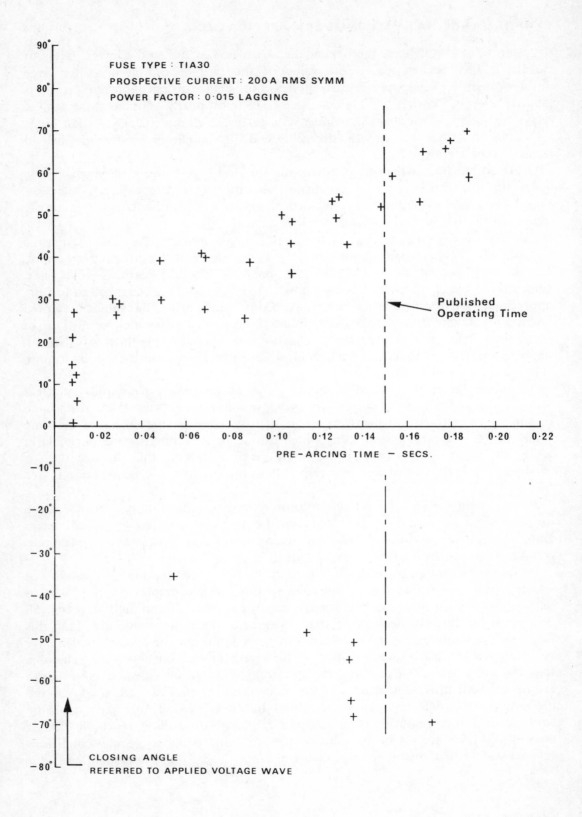

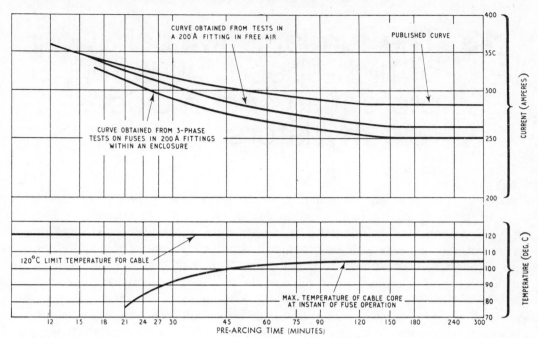

FIG. 12.13

Effect of ambient temperature and other environmental effects on fuse characteristics. Curves show that effect is insignificant for times shorter than about one hour

persisted in. Conventions such as this have been used in the past for choosing fuses for the protection of semiconductor devices, with less than satisfactory results. Another factor to be taken into account is the relationship between the applied voltage and the inherent maximum voltage capability of the fuse in question. A fuse can often successfully interrupt a circuit where it is used to the limit of its voltage capability, but in such cases the arcing $I^2 t$ can be expected to be very many times the value of the pre-arcing $I^2 t$.

FIG. 12.12

Variation in pre-arcing time with closing angle. Record of actual test showing variation of pre-arcing time due to effect of random switching at inductive moderate overcurrents. Note that predominance of test shots on rising voltage is fortuitous and due to method of actuating making switch. Fuse type, TIA30. Prospective current, 200 A r.m.s. symm. power factor, 0.015 lagging

13 The fundamental behaviour of high-speed fuses for protecting silicon diodes and thyristors

As the development and application of semiconducting devices proceeded the demand for more economical loading reduced the margin for protection. The main problem was that of defining the fault withstand of the devices in more precise terms and also in terms which were directly comparable with those applying to the fuses used to protect them. It is necessary to ensure that the fault energy 'let-through' of the fuse comes within the withstand of the device in those circumstances for which the fuse is designed to cater. Both must be expressed in like terms. Problems arose in defining the behaviour of semiconductor systems under transient conditions during the extremely short operating time of their associated fuses.

An invitation to address the I.E.E.E. International Convention in New York in 1968 provided an opportunity to raise the issue and to provide a forum for fuse and device designers to discuss possible solutions.

It should not be supposed that because the fuse appears to be simple in construction that it is less complex in thermal terms than the semiconducting device. The problem can perhaps be seen by considering the fact that the fuse must change its state completely when clearing the fault while the device remains unaltered and undamaged. The range of thermal processes affecting the one are very different to those affecting the other and of the two, the fuse processes are the more complicated.

Although there is considerable improvement in understanding between semi-conductor and fuse information, this issue will continue as a challenge to designers for as long as devices continue to develop. By its nature, fuse development must always follow the development of the equipment it protects. The alacrity with which fuse designers can meet the needs of newly developed equipment once they are known often decides how quickly new equipment can be introduced on to the market.

Originally published by The English Electric Co. Ltd., Fusegear Division in 1968

13.1 Introduction

The fundamental problem in applying fuses for the protection of semiconductor devices is that of ensuring that the fault energy let-through of the fuse comes within the withstand of the device under all the circumstances which can arise in service. This presupposes that the behaviour of both fuse and the device are known in like terms over the whole range of duties likely to be encountered. Also that the circumstances which occur in service can be predicted with reasonable certainty. Protection can only be applied to the extent that the essential parameters and conditions to be met can be identified and specified in properly related terms. The degree of protection achievable depends not only upon the performance of the fuse and of the device it protects, but also upon the degree to which the performance can be expressed in terms which are mutually meaningful.

The parameters on which semiconductor withstand is normally compared to fuse let-through are peak current; $I^2 t$ let-through; rate of rise of current (di/dt); and peak voltage. All these parameters other than di/dt are functions of the mode of operation of the fuse as well as of the circuit. Individual circuits are subject to a great variety of conditions which are in turn imposed upon the fuse. Particular fuses can also be applied in a great variety of circuits. Other variables are introduced in service, due to environmental conditions particularly those affecting temperature and other thermal effects. The fuse responds to these conditions in different ways, depending on the combination of circumstances which apply at the time.

di/dt is wholly controlled by the circuit because it reaches its greatest values before the fuse melts. The rates of change of current which occur during the arcing period of the fuse are not serious unless they produce dangerous arc voltages [$L(di/dt)$] but these can easily be controlled by good design.

The mode of operation of the fuse when interrupting a fault does not follow a simple time/current law. This introduces some complication in presenting fuse performance data, but it also provides the very means by which fuses are able to interrupt rapidly enough to limit the fault let-through and so provide high-speed protection.

Fuse performance can be ascertained over a very wide range of circuit conditions by empirical means or by calculation. It is then necessary to decide upon the most effective method of presenting performance data to suit the needs of particular users. The parameters which predominate for some applications may be less important for others and the number of combinations of circuit and fuse parameters can be very large. This is why it is necessary to adopt conventional means of presentation of data and to accept a degree of rationalization. The same arguments apply equally to the presentation of withstand data of semiconductors and the issues which exercise the minds of both semiconductor and fuse designers are those concerning the necessary reconciliation between fuse data on the one hand and diode/thyristor withstand data on the other. It is necessary to recognize that the thermal and other characteristics of these two devices are different by nature. Successful protection under one set of conditions cannot be taken as necessarily implying that satisfactory results will be obtained under another.

The factors used for defining withstand and let-through must be seen in proper context. For instance Ip and $I^2 t$ are useful as comparative factors, but they do not quantify energy or stress in absolute terms. The condition for protection is that the fuse element must get very hot so as to change its state and interrupt the circuit, while the semiconductor must remain sufficiently cool to avoid any change of state whatsoever. Thus, the energy and stress absorbed by each must necessarily be different, in both magnitude and degree. notwithstanding the fact that they both experience the same let-through (as it is presently defined). Moreover, the stresses experienced by each may vary at greatly different rates under different conditions.

These differences are no more and no less than laws of nature which have to be understood in attempting to bridge the gap between diode/thyristor behaviour and fuse performance. High-speed fuses have been successfully protecting diodes and thyristors for many years, which is evidence enough that the knowledge which exists has so far been adequate. But progress constantly demands new information at a substantial rate. There is still very considerable potential for fuse development in this field.

13.2 Balance of parameters

The choice of a fuse for optimum results requires a proper balance of all the parameters involved. All parameters are closely interrelated and cannot function in isolation. A concentration of attention on one parameter must not ignore its relationship with the others.

Fuses can be made in a large variety of current and voltage ratings with a variety of characteristics corresponding to each. Degrees of time grading, of peak current and I^2t let-through, arc voltages and interrupting capacity are all achievable. Interrupting capacity is sometimes taken for granted because it does not often appear as the limiting factor to the modern fuse designer. Nevertheless, pre-occupation with the other parameters should not obscure the fact that interrupting capacity is still the primary property to be considered. Mathematical extrapolation of the interrupting capacity curve is no substitute for proving fuses at full power up to the maximum available current for which rating is claimed.

A balance of parameters which is appropriate to one application may not be appropriate to another. This is why high-speed fuses specifically designed for protecting semiconductor devices have come into being. For this purpose, emphasis must necessarily be upon those parameters which affect the more sensitive areas of semiconductor behaviour. Generally speaking, the emphasis centres around the control of let-through and arc voltage in relation to the higher orders of short-circuit current rather than to the zones corresponding to the more moderate fault currents. This is because the nature of the faults which occur when a semiconductor device fails is such as to produce substantial values of available current.

Fuses have to be designed to adequately perform both passive and active functions. These are mutually incompatible from a purely fundamental point of view. The passive function is to carry load currents without exceeding a given working temperature. The active function is to interrupt the fault current as quickly as possible. The passive function must include the ability to withstand transient overloads as well as steady load currents. The active function must include current and energy limitation and the ability to absorb such inductive energy as may be present in the circuit at the time of interruption. The ultimate objective in fuse design is, therefore, to produce a fuse having minimum power loss associated with the passive function and maximum energy limitation with minimum arc voltage associated with the active function.

A knowledge of how the essential parameters are interrelated is necessary in judging the overall effectiveness of a particular design for a particular application. For instance, it is possible to obtain a minimum value of $I^2 t$ let-through by designing the fuse to chop the arc fairly rapidly. This may incur the hazard of producing a higher arc voltage than would be produced by a fuse designed to dispose of the arc in a more uniform manner. The former fuse would also incur a higher watts loss and run hotter when carrying normal load. To remedy this would mean a strengthening of the fuse element which in turn would increase the let-through and thus tend to defeat the original objective. A better approach to the problem would be to design the fuse element to have better thermal characteristics. In other words to have more efficient heat transfer from the fuse element.

It is not necessary to design a special fuse for every individual circumstance because it is possible to use a given fuse in a variety of circuit conditions providing its behaviour under all conditions can be predicted. A simple, but very typical, example of this is that of applying a fuse at a much lower value of voltage than its rating in order to achieve a lower value of $I^2 t$ than would be obtained by a fuse having a rated voltage similar to the applied voltage. This is possible for those designs in which the arc voltage (which is usually a limiting factor) keeps in step with the low applied voltage, as illustrated in Fig. 13.1.

There are numerous other such examples. In fact, the number of combinations which can be achieved by manipulating the various parameters is considerable. This denotes versatility rather than complexity in fuse application and should be seen as such. The fact is that although fuses are becoming very sophisticated in design they are still, relatively speaking, simple as hardware. They react in different ways to conditions imposed upon them and it is usually the conditions which make the software complex. The use of computer terminology is not inappropriate.

13.3 Current rating

The natural current rating of a fuse is a value less than minimum fusing current which the fuse can carry continuously without deterioration or change of characteristics, Current rating is usually assigned in relation to a limiting temperature rise and conventionally chosen environmental conditions. The fuse, being a generator of heat, must dissipate heat by conduction to its associated connections and mounting and by convection and radiation from its surfaces. Current rating outside the context of these conditions has no meaning — a fuse cannot be rated in

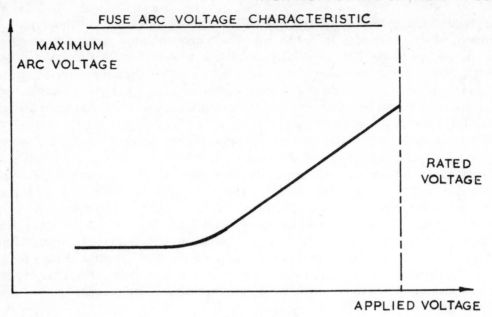

FIG. 13.1
Fuse arc voltage characteristic

isolation. The aim of good design is to keep the watts loss to a minimum consistent with proper performance of the active functions and much can be done in this direction (at reasonable cost).

In semiconductor equipments it is necessary to relate the conventional basis of rating of the fuse to the actual condition obtaining when the equipment is in use. Some correction to the current rating may have to be applied if the fuse is associated with conductors and ambient air which run hotter than those which apply to the basis of rating. General rules governing such correction are difficult to formulate because fuses and service conditions are widely variable. Usually, it is sufficient to state the maximum temperature at which a fuse may be used and to apply a rule of thumb for derating when this is necessary. In many cases the load factor is such that it compensates for the higher environmental conditions and derating is not necessary. The effects of natural or forced cooling or of water cooling of conductors, etc., need to be included into the equation.

The maximum temperature which a fuse will withstand obviously depends upon the materials used in its construction. The choice of materials may be dictated more by the requirements of the active function than of the passive function. If an element structure and its joints are made from high-temperature materials, they may withstand severe load conditions, but may not suit the other materials which form the fuse package as a whole. The danger zone in this case may well be that in the region of minimum fusing current which can produce temperature rises well beyond the withstand of the fuse tube. (Temperature is proportional to something more than the square of the current and the temperature in question is that of the fuse element.) The other factors are the rate of heat transfer from the element to the tube and the rate of dissipation from the tube.

Users cannot be expected to concern themselves with the intricacies of such thermal problems. Thus, the conventional current rating of a fuse must always be chosen so as to allow margin for these contingencies and to allow for transient overloads such as switching surges and other similar occurrences. The short-time rating of a fuse is governed by its time/current characteristic. It is necessary to know how close to the time/current curve a fuse can be worked without incurring deterioration. The margin allowable varies according to the design. A simple homogeneous element differs in this respect from a heterogeneous design made up by alloying several metals. It is not only a matter of working temperature, but of thermal mass or inertia. A fuse has a thermal memory which is useful when properly used, but can be a nuisance if overlooked. Mechanical stresses due to thermal cycling and metal fatigue are also factors which have to be accounted for in design. One of the most trying duties of a fuse is that of withstanding repeated current pulsing. This capability can be provided by design – the difficult problem is usually that of reliably estimating the true magnitude and nature of the pulses. It is often found that fuses are more sensitive to pulses than the instruments used to measure them. This means that the estimates of loading are sometimes fictitious and often unwittingly so.

Chemical compatibility between materials used in fuse construction to avoid long-term deterioration is also a relevant factor.

From the user's point of view, non-deterioration is an essential prerequisite to safe protection. Fuse characteristics cannot be checked after installation by any non-destructive means. If they change they cannot be recalibrated, nor should it be necessary that they should. Fuses should be designed, rated and applied so that they can remain unchanged without attention over very many years of service. Safety requires that they should be as effective at the end of their lives, which may be twenty years or more, as at the beginning.

13.4 Voltage rating

Voltage rating is also a conventionally chosen value which is lower than the maximum value at which a fuse will safely interrupt a circuit under defined conditions. A given fuse can have several voltage ratings according to the manner in which it is intended to be used. It does not follow that a fuse is safe at all applied voltages below its rating irrespective of circuit conditions. Normally, it will have been designed for one voltage rating relating to a particular range of duties. If used outside the range of duties for which it is intended, then the voltage rating requires modification to suit the alternative conditions.

An obvious example of this principle is the comparison between the a.c. (60 Hz) rating and the d.c. rating. For circuits of moderate inductance the d.c. voltage rating is usually something less than half the crest voltage associated with a.c. rating, depending upon the design of the fuse and the margins inherent in the rating. The voltage capability of a d.c. fuse may be different for the high available currents than for the lower currents. This rule applies also to current-limiting fuses used for high voltage. On fuses which are expected to include all orders of available current the nominal voltage rating (marked upon the fuse) must obviously be the safe value covering the zone of currents to which the fuse is most sensitive.

At low available currents a fuse element melts relatively slowly into a single arcing condition. At high available currents simultaneous multiple arcing occurs. A convenient dividing line between the two is the threshold at which current limitation or cut-off becomes apparent.

With d.c. the single arcing condition is the more difficult one to interrupt. With a.c. the single arcing condition becomes a problem only at the higher voltages. Voltage rating is usually related to interrupting capability, but where degrees of current and energy limitation are the primary requirement voltage rating must necessarily be related to these at values lower than those which correspond to the intrinsic maximum interrupting capability.

Another factor which influences voltage rating is that which governs the ratio between melting $I^2 t$ and arcing $I^2 t$. This can be important for co-ordination in the circuit. If a fuse is designed for a very low value of melting $I^2 t$ and is applied at a voltage close to its inherent maximum interrupting capability, the arcing $I^2 t$ will be relatively large. In extreme cases it is not unusual to find that the arcing $I^2 t$ can be 10 to 20 times greater than the melting or pre-arcing $I^2 t$. This usually involves in addition a significant rise of current after arcing has commenced such that the peak current let-through may be several times the value of current at the commencement of arcing. Although under the circumstances the fuse may interrupt safely, the disproportion between pre-arcing and arcing $I^2 t$ is so contrary to the conventions which users have come to expect that it may give rise to dangerous situations if not recognized. Closer ratios are easily possible, but a fixed ratio of 1 : 2 adopted in some publications does not accord with fact. It is more important to choose an optimum melting $I^2 t$ which although it is not the minimum achievable so co-ordinates with the voltage rating as to give a minimum total $I^2 t$. This automatically gives a favourable ratio of melting to arcing.

Inductance is another factor which influences voltage capability and rating. Generally speaking the higher the circuit inductance, the lower the voltage capability. Power factor is significant to voltage rating in the case of a.c. fuses, but as this is normally chosen conventionally on the basis of a low value (about 0.15) any reduction in power factor beyond this has little consequence The effect of time-constant is equally significant in the case of d.c. fuses and may apply, over a wider band of time than in the a.c. case, because this is limited to a time determined by the zero pauses.

As a general rule it is necessary to differentiate between the various applications and to recognize the sensitive areas inherent in each. It is then possible to choose a fuse with appropriate voltage capability and to have concern for those conditions which may be expected to arise. For instance, fuses used for back-up purposes are not expected to experience low overload conditions and may, therefore, be rated and applied without regard to them.

It is obvious that the voltage capability of a fuse is not a single value, although in the interests of rationalization it may be marked with a particular rating relative to a recognized specification. Such rationalization is useful for rule-of-thumb working, provided it does not obscure the many other possibilities which fuses can satisfy.

13.5 Time/current characteristics

The inverse time/current characteristic which fuses obey can be varied very considerably by design. As with all other manufactured products the fidelity of fuses to the time/current characteristics claimed for them is dependent upon the tolerances to which they are manufactured. The achievable accuracies inherent in the fuses themselves are of high order and the best designs can be made to be easily consistent with the semiconductor devices which they are employed to protect. Apart from

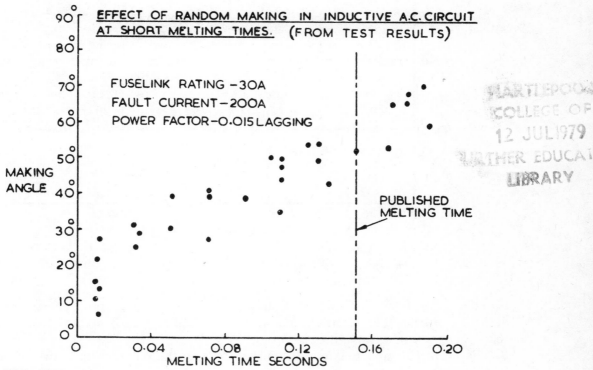

FIG. 13.2

Effect of random making in inductive a.c. circuit at short melting times (from test results). Fuse-link rating, 30 A; fault current, 200 A; power factor, 0.015 lagging

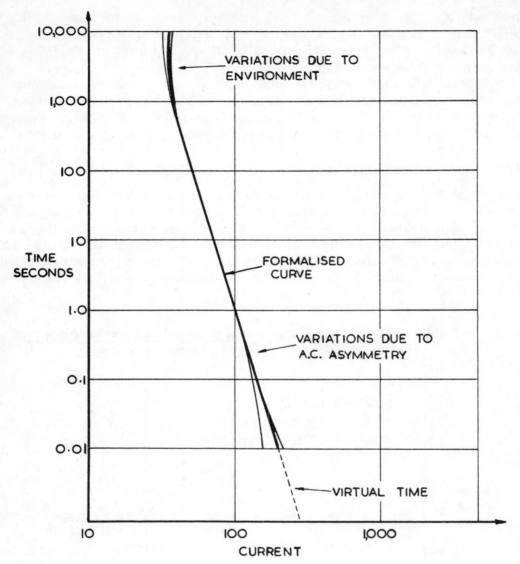

FIG. 13.3
Time/current characteristic

manufacturing tolerances the variations which occur in time/current characteristics derive more from the factors external to the fuse and to the method of presentation than to any peculiarities in the fuse itself.

It is normal to plot time/current curves showing the relationship between r.m.s. symmetrical available current and pre-arcing or melting time in the zone between minimum fusing current and the threshold of cut-off. For a.c. the total operating time is not significantly different from the melting time because arcing seldom persists for more than a fraction of a cycle and is usually negligible compared to pre-arcing time. The same rule applies to d.c. only if fuses have a voltage rating appropriate to performance at the lower orders of available current.

In the low overload zone approaching minimum fusing current a fuse is susceptible to environmental effects. These are usually not in evidence except for operating times exceeding several minutes. If a fuse is properly chosen as regards current rating, it is unlikely that variations from the time/current curve in this zone will be significant in the protection of semiconductor devices.

At the short-time end of the curve, for operating times in the region of one second or less, the incidence of the making or closing angle in relation to the circuit constants can cause considerable scatter in the time/current points. This sometimes causes confusion where co-ordination between fuses and other devices is attempted on the basis of time grading. In the immediate region of cut-off threshold pre-arcing time can vary over a ratio of more than 20:1 due to variations in making angle, as shown in Fig. 13.2. This is the natural response of the fuse to the various rates of rise of current occasioned by current asymmetry.

Conventional time/current curves cannot show these variations because they are formalized to show only the time related to r.m.s. symmetrical current. Published time/current curves sometimes show pre-arcing times which actually refer to a degree of asymmetry even though they are formally plotted against r.m s. symmetrical values. This is done to reflect the practical conditions of service. If co-ordination between fuses and other protective devices and the equipment they protect is made on the basis of such curves, then time as an independent factor and variations due to asymmetry can be ignored for practical purposes. Fig. 13.3 indicates some of the variations involved.

The variations due to asymmetry in a.c. circuits have their counterpart in d.c. circuits due to inductance and time constant. Since the common factor between both a.c. and d.c. is energy required to melt the fuse element the two time/current characteristics obviously coincide for the longer times at which steady-state currents have been reached. It is also possible to show a relationship between them for the shorter times and to represent them graphically as in Fig. 13.4 so that d.c. characteristics can be derived from a.c. curves and vice versa.

Beyond the threshold of cut-off the time/current relationship changes drastically and for this zone other forms of presentation have to be adopted. But time/current curves can be extended into this zone by mathematical artifice. One such method is the adoption of the concept of 'virtual time' which is written into a number of national and international fuse specifications. This method simply presents $I^2 t$ let-through on a basis of r.m.s. symmetrical current which results in a 'virtual time':

$$t_v = \frac{i^2 \, dt}{I^2 \, \text{rms (symm)}}.$$

If it is assumed that pre-arcing $I^2 t$ is a constant over a particular band of times, then virtual time becomes a linear function which can be determined from two

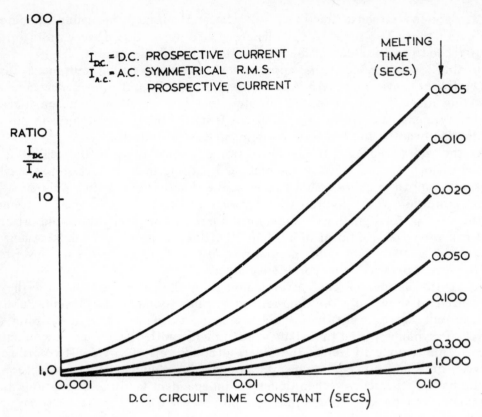

FIG. 13.4
Curves used to determine short-time d.c. characteristics from existing a.c.
time/current data I_{DC} = *d.c. prospective current;* I_{AC} = *a.c. symmetrical*
r.m.s. prospective current.

points. Virtual time can also be applied to total let-through if properly qualified to account for variations of asymmetry and applied voltage.

13.6 I_p and $I^2 t$

Primarily, I_p is a function of melting $I^2 t$ and the average rate of rise of current throughout the melting time, the one being a function of the fuse and the other a function of the circuit. There is no evidence of any limit to the rate of rise of current to which a fuse will respond. (Tests on fuses protecting a large capacitor bank showed them to exhibit cut-off when clearing an available-current of more than 200 kA (peak) at 3000 Hz. This represents $di/dt = 2 \times 10^9$ amp /sec which is a value greater than that currently quoted as the limit of capability of semiconductor devices.)

The energy required to melt a fuse element consists of the heat required to melt the element (including the latent heat of fusion) plus loss during the melting time.

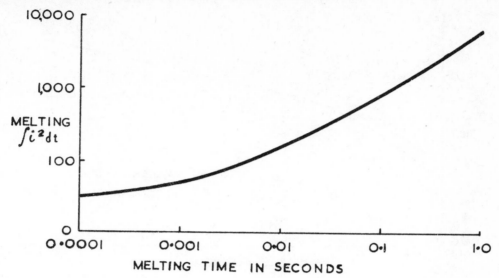

FIG. 13.5
Fuse melting $\int i^2 dt$ characteristic of typical fuse

The loss component remains significant down to very short times, but the actual time at which it can be ignored for practical purposes varies largely according to fuse design. In the case of high-speed fuses the reduction of melting $I^2 t$ with the shorter times becomes especially relevant.

When presenting I_p it is necessary to simplify and formalize the published information. Wisdom may dictate that the values published should be those

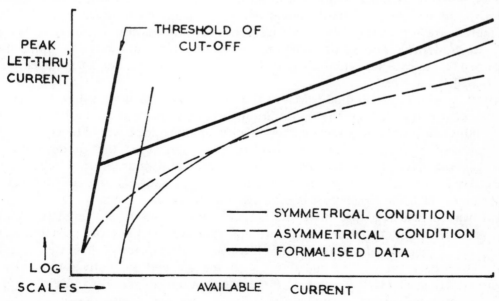

FIG. 13.6
Example of formalized data representing 'worst case' condition

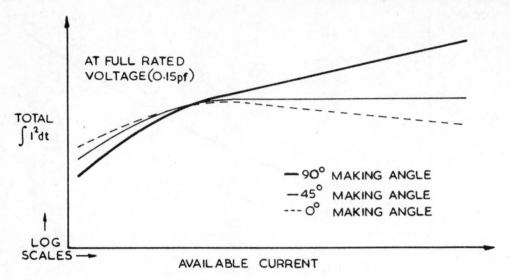

FIG. 13.7
Illustrating 'worst case' in respect of making angle

representing the most onerous condition which can arise in service. Mitigation can then be applied where less onerous conditions are known to exist. For instance the rate of rise of current is greater in the symmetrical condition than the asymmetrical and I_p correspondingly greater (see Fig. 13.6). On the other hand, the threshold of cut-off (referred to the r.m.s. symmetrical available current) occurs earlier with the asymmetrical condition and needs to be known.

It is well known that arcing $I^2 t$ is proportional to I_p (and thus to available current) and to applied voltage, power factor and making (or arcing) angle. These parameters are not always obvious from published information. The safe course is again to show the results of the most onerous combination, but in this case it is also necessary to indicate variations at least due to applied voltage and available current (see Figs. 13.7 and 13.8).

It can be argued that in diode circuits, the breakdown of a diode will occur during the blocking period so that the fault current becomes apparent at the beginning of the conducting period. If this were true, then the making angle would be constant in the region of 0° (voltage wave). Practice has shown, however, that diodes fail in a more random manner and the making angle can be any value. Accordingly, diode fuses should be tested at various closing angles to ascertain the maximum $I^2 t$ let-through which the circuit/fuse combination can produce.

The way in which a fuse interrupts an a.c. arc is necessarily different from the way in which it interrupts a d.c. arc because the average value of applied voltage during the arcing period is different in either case. Consequently, there is no simple relationship between a.c. $I^2 t$ and d.c. $I^2 t$. If the a.c. and d.c. voltage ratings of a given fuse are such that the value of $I^2 t$ in the region of the cut-off threshold is the same for both voltages, it is unlikely that d.c. $I^2 t$ for higher available currents will

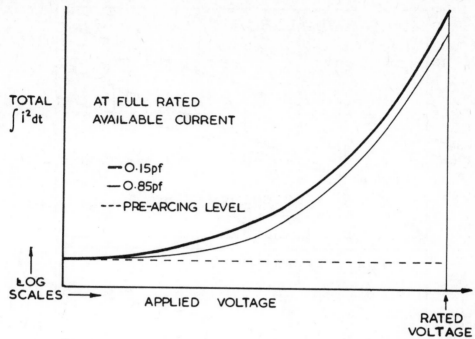

TOTAL ∫ i²dt

AT FULL RATED
AVAILABLE CURRENT

— O·15pf
— O·85pf
--- PRE-ARCING LEVEL

LOG
SCALES ——→

APPLIED VOLTAGE

RATED
VOLTAGE

FIG. 13.8
Illustrating 'worst case' in respect of power factor

exceed the a.c. values. The d.c. $I^2 t$ is likely to remain at a more constant value than the a.c. as shown in Fig. 13.9. This supports the view (already proved by test and in practice) that the upper limit of interrupting capacity can be very high for d.c. fuses and much beyond the values likely to be encountered in service. It also indicates that the sensitive areas for d.c. fuses are those involving the lower orders of fault current, particularly those with longer time constants.

The precise relationship between a.c. and d.c. in this respect depends upon design and requires more study. Typical current and voltage oscillograms, as in Fig. 13.10 for a given fuse at differnt values of available current, clearly show the trend and indicate the possibilities.

The rate at which a fuse cools the arc and builds up its dielectric strength is reflected in the arc voltage. One of the advantages of a fuse is that unlike other forms of circuit interruption, the rate of interruption can be finely controlled. This means that both the inductive components $[L(di/dt)]$ and the arc drop can be controlled to the same degree. The fuse provides its own damping mechanism. Whether the mode of interruption of the arc is seen as a cooling of it or as a suppression by the creation of arc voltage is a matter of choice. Within the limits of arc voltage imposed by the system insulation (or semiconductor withstand) arcing $I^2 t$ must obviously have a direct relation to arc voltage. Where $I^2 t$ withstand is at a premium there is obvious virtue in accepting as high a value of arc voltage as the equipment will permit.

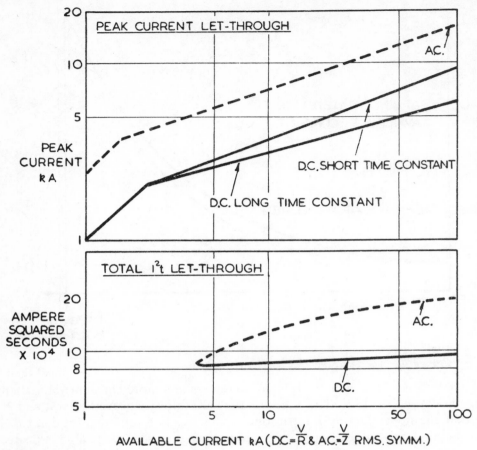

FIG. 13.9

Comparison of let-through values for a.c. and d.c. test conditions on a given fuse with typically assigned voltage ratings

13.7 Other d.c. factors

One fact which emerges strongly from the comparison of a.c. and d.c. performance is that it is safer to see them as separate entities. Relationships between the two are tenuous and in some areas defy calculation, this means that each must be considered on its own merit and on its own empirical basis, Figs. 13.11 and 13.12 are typical of these.

Most diodes and the fuses protecting them in rectifier circuits see what are essentially a.c. conditions, thyristors used in inverter circuits see d.c. conditions. Moreover, thyristors require more sensitive protection and are subject to hazards which do not normally beset simple diodes. A gating fault in a three-phase inverter coincident with a thyristor failure can impose upon the protecting fuse a voltage of

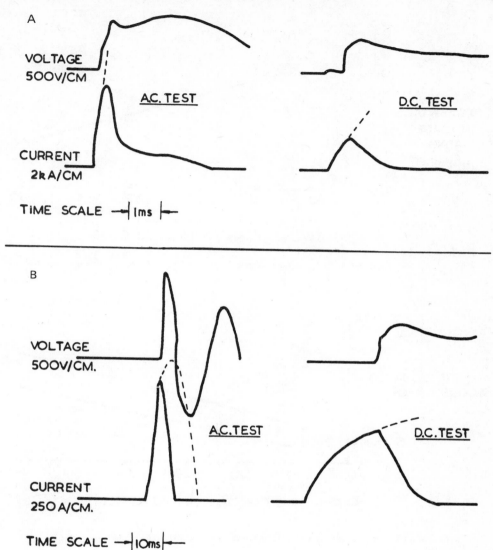

FIG. 13.10

Comparison of a.c. and d.c. behaviour of a typical fuse for thyristor protection: (a) at high available currents; (b) at available currents in region of cut-off threshold. d.c. voltage rating ≑50% of a.c. (P.I.V.) rating

twice that of the voltage rating of the thyristor which it protects. Another hazard which may affect rectifier equipment is that of d.c. feedback from common d.c. busbars. This can occur when several equipments feed into one common busbar, and if a fuse is expected to deal with feedback of this sort it must be designed and rated to cope with composite a.c./d.c. faults of specified magnitude.

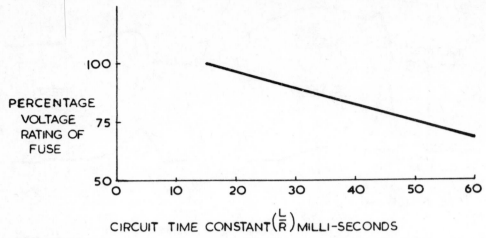

FIG. 13.11
Effect of time constant on d.c. voltage rating. Referred to threshold region of available current

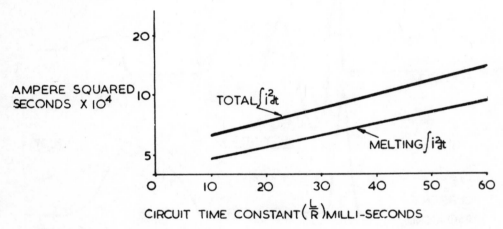

FIG. 13.12
Variation of $\int i^2 dt$ with circuit time constant at constant circuit voltage and available current

13.8 Frequency and non-sinusoidal waveforms

It has been shown that a fuse designed and rated for one frequency will, generally speaking, safely interrupt currents of the same magnitude at a higher frequency. Many investigations have been done at between 50 and 60 Hz. The peak current is marginally higher at the higher frequency due to the greater rate of rise of current, but the I^2t is marginally less. Since I^2t is more relevant as regards stresses which affect the fuse, the higher frequency is almost always less onerous than the lower.

Tests have shown this rule to be true for ascending values of frequency. Melting I^2t is reduced as a result of the reduction in melting time, the relative peak current is

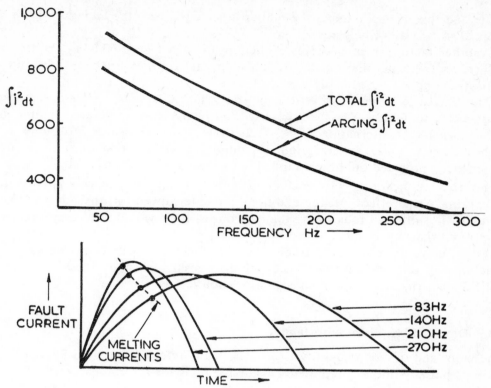

FIG. 13.13
Variation of $\int i^2 dt$ let-through with frequency (from test oscillograms)

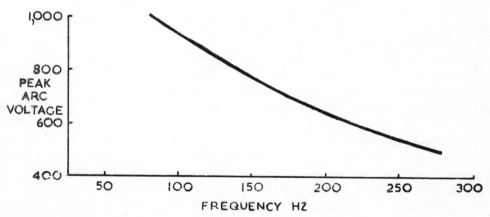

FIG. 13.14
Variation of arc voltage with frequency for a typical fuse

lower and this in turn reduces the inductive energy. Thus, the arcing $I^2 t$ is less and thermal stresses as a whole are reduced.

Another factor which becomes evident from such tests is that arc voltage also reduces as frequency increases. The basis for this requires further study, but the mitigation implied is a valuable asset.

The increasing use of thyristor inverters introduces an increasing range of frequencies to which fuses are subjected. In many cases these are unidirectional pulses rather than alternating waves. Tests show that when a fuse interrupts such a high-frequency pulse, interruption is completed at the first zero after arcing commences, provided the fuse is applied at its proper d.c. voltage rating.

Inverter applications also include a great variety of non-sinusoidal voltage waveforms. The effect of these during arc interruption is difficult to predict, although it is difficult to conceive that any one of them would impose more onerous conditions than a pure d.c.

Fuses used in rotating rectifiers may see low-frequency components due to pole-slipping and it is usually necessary to consider each of these cases on its individual merits in arriving at an economic choice.

13.9 Diode and thyristor withstand

The withstand of diodes and thyristors is normally referred to a sinusoidal half-cycle (60 Hz) pulse. While this represents a suitable basis for $I^2 t$ comparison, it does not necessarily indicate the peak current or the short-time pulse withstand. Various conventions have, therefore, had to be evolved to obtain expressions which are mutually meaningful to diode/thyristors and fuses.

It would appear that a convenient method of determining diode/thyristor withstand is by type (design) testing with sinusoidal half-wave pulses of varying frequency. In the case of a diode used in a 60 cycle rectifier it may then be possible to refer the $I^2 t$ to the 60 cycle pulse and to obtain I_p from a pulse of higher frequency (e.g. 250 Hz). The limiting factor for the higher frequency may be the limit of di/dt which the diode can withstand.

In cases where the applied frequency is greater than 60 Hz it is to be expected that $I^2 t$ withstand will diminish. Fortunately, the same thermal rules apply in a general way to the fuse, but it is necessary to ensure that the thermal characteristics of the diode/thyristor are in fact compatible with the fuse characteristics in this respect.

A further problem which needs to be resolved is that which arises from the fact that the current waveshape let through by the fuse deviates considerably from the sinusoidal. It is thus necessary to find some convention by which the fuse let-through can be expressed in equivalent sinusoidal terms or, alternatively, to express diode/thyristor withstand in terms which are more directly comparable with fuse behaviour. It would appear that the former is the more logical, although suggestions have been made that conversion of both withstand and let-through to equivalent triangular or square wave shapes might also be appropriate. By the nature of the problem it is not possible to regularize the fuse let-through in service as this is a natural function, determined by the fuse/circuit combination. On the other hand, the formal waveshape adopted for the determination of withstand can be arranged to simulate any regular pattern. Mutual agreement between semiconductor and fuse designers is needed, because it would be an obvious advantage if conventions could be evolved which are universally applicable.

The fuse is a relatively simple piece of hardware which can be made to respond with fidelity to a large number of combinations of complex conditions. This is a measure of its versatility. The rationalization which is necessary to reduce the complexity of conditions should, for safety, represent the most onerous duty, but should not obscure versatility by oversimplification.

14 Evaluation of reliability: H.R.C. Fuse-links

The new science of reliability evaluation has emerged as a 'spin-off' from the U.S. space programme. The proposition put forward is that it should be possible to evaluate the reliability on a common basis of every component in a complex system such that the whole may be expected to function reliably in service. The science has advanced to a point where it is the preoccupation of many university faculties and is now invoked as a 'condition of contract' for complex pieces of equipment in industry.

Fuses appear as components in many equipments, from battleships to rockets and from computers to nuclear power stations. Hence, when a meeting of specialists on the reliability of electrical systems in nuclear power stations met in 1968 to consider aspects of reliability evaluation, a request was made for the presentation of a paper to consider fuses in this context. An investigation was carried out according to the laid down evaluation criteria and was duly reported. Specific results were obtained and it was seen that these had much wider implications as further proof of the reliability of fuses in industry as a whole. (It should be recognized that the results obtained applied to a particular design and make of fuse.) The paper outlined the principles upon which reliability may be evaluated and showed that it can be specified and provided to a required degree.

Presented to the European Nuclear Energy Agency Committee on Reactor Safety Technology at Ispra, Italy in June 1968 and originally published by The English Electric Co. Ltd., Fusegear Division in 1968.

14.1 Introduction

Reliability must obviously refer to the overall performance of a given piece of equipment and in this respect protective devices are in a somewhat different category to those more static pieces of electrical equipment which are only expected to function satisfactorily under normal conditions. The ability to perform satisfactorily over a period of time in this way under reasonably normal conditions is a passive function. It usually includes, within economic limits, some inbuilt safety margin in the form of overcapacity to cope with the various forms of temporary stress which may occur, but beyond this the equipment would be expected to fail.

Such equipment and the systems in which they are used are usually protected by other devices which absorb those stresses liable to occur under abnormal conditions and effectively disconnect the faulty circuit. Thus, protective devices must perform both passive and active functions; those of carrying current under normal conditions, and those of interrupting it when abnormalities occur. The two functions are physically interdependent in as much as any deterioration taking place during the passive life may impair the ability of the device to perform its active function effectively when required to do so.

It is evident that fuses must be seen in the context in which they are used, and some knowledge of the technology which governs their design and performance understood before reliability in this context can be evaluated to any meaningful degree.

This chapter is, therefore, written in two parts:

(1) An objective, though rather general, review of those aspects of fuse technology relating to design, manufacture and usage.

(2) A subjective account of the methods used to maintain the standards of quality and reliability of 'English Electric' fuses, followed by a quantitive first approximation of reliability by accepted evaluation methods.

14.2 Fuse technology and usage

The primary function of an electrical system is to deliver electrical energy to where it is wanted. If it could be relied upon not to fail, protective devices would not be necessary. The first line of protection, therefore, lies in the inherent quality of the components which comprise the system. Prevention is the best form of protection and this can be achieved to a desired degree by attention to planning, correct choice of materials and proper installation appropriate to a particular purpose and situation. When everything has been done to ensure that the system is well founded, protective devices, and fuses in particular, have application as insurance against those unavoidable faults, which by their nature cannot be predicted or prevented.

Fuses and other protective devices are best regarded as necessary appendages to the system rather than as primary components. They ensure the continuity of the service, but should never be used to shore up an otherwise bad installation, except as a temporary expediency with full regard to the risks involved.

In practice this counsel of perfection is seldom observed because of the enormous diversity both in attitudes among users, and in the conditions under which fuses work. Any method of evaluating the reliability of H.R.C. fuses as such must necessarily take into account the state of knowledge of the people who use fuses and make due allowance for the various degrees of misapplication which inevitably occur. Fuse designers and those who write fuse standards have in the past recognized this state of affairs and have allowed fairly adequate safety margins in their specifications on the assumption that fuses are to a considerable extent handled by people who do not necessarily appreciate even the fundamental bases of fuse technology.

Fuses must respond to any electrical disturbance which may be damaging to the system and operate before there is time for damage to occur. At the same time fuses must not by their presence weaken the system or restrict its ability to perform its primary function in any way. This means that fuses must be intrinsically more reliable than the systems which they protect: a condition which imposes an apparent contradiction in terms which needs to be resolved before a fundamental concept of fuse reliability can be properly defined. In common parlance, the fuse must be stronger in its passive function than the other components in the system which it protects, but must be weaker during its active function. Thus, reliability in this sense depends upon the degree to which 'strength' can be maintained on the one hand and 'weakness' controlled on the other.

The purpose of fuses is to remove individual faulty circuits as quickly as possible so that the rest of the system can continue to function without impediment. Generally speaking, fuses exist to protect the system rather than the circuit or the apparatus in the circuit, although there are circumstances under which they may do so. They are short-circuit rather than overload devices, and although they will give a reasonable measure of overload protection and will protect apparatus with reasonably long thermal time constants such as cables, against overloads, they should not be expected to protect apparatus with short-time ratings unless specially designed for this purpose.

The most compelling testimony to the success of fuses in reliably providing safety over the last thirty years is the enormous growth in their usage and the high reputation which certain designs have earned and maintained. Electrical services throughout industry in the U.K. and Commonwealth countries and elsewhere in the world are almost completely dependent upon H.R.C. fuses, particularly in the low- and medium-voltage field. This reputation rests much more upon technical achievement in design, manufacture and application than is apparent in the seemingly simple construction of the fuses themselves. Although fuses are simple in conception and derive reliability from the simplicity of their construction, they are exceedingly complex in function in as much as they must respond to the complexities of any system which they are called upon to protect. Systems and applications are almost infinite in variety and in spite of attempts to standardize, the variety of fuses available and the rate at which new designs are produced is still very great.

14.2.1 *Performance parameters*

In practice a fuse performs its passive function of carrying normal load currents during its service life, which may be anything from minutes to years, up to the time it performs its active function of interrupting a fault. The incidence of the fault cannot be predicted and therefore, for design purposes, the life expectancy of the fuse must be assumed to exceed the life of the system in which it is installed. Many systems continue in operation for thirty years or even more, and the fuse must be able to deal effectively with any magnitude of fault which occurs at any instant during this period.

During its life the fuse carries load currents which may vary from zero to full load with varying time cycles giving rise to thermal changes with resultant mechanical effects. The fuse must be capable of withstanding such transient overloads and through-fault currents as may be expected in normal service. These may introduce electromagnetic stresses concurrent with the thermal and other stresses, all of which must be absorbed by the fuse without incurring permanent change. Conventionally the ability of the fuse to carry current is determined on the basis of a limiting temperature related to the full-load condition. The aim of the designer is to achieve as low a temperature rise as possible for this condition, so as to minimize resulting strain.

The manner in which fault currents are interrupted depends upon the transient behaviour of the system as well as upon the mode of operation of the fuse itself. The fuse element melts in a time which is basically proportional to the square of the current, but the value of current is variable, being itself dependent upon the system constants. Thus, it is the current integral with respect to time which is the relevant value. During the very short times which are involved in interrupting heavy faults, the complexity of the thermal processes of melting have to be understood in detail in order to achieve predictable results. When the element has melted, the energy in the circuit creates an arc which has to be contained, dispersed and cooled to complete the process of interruption. The fuse must be capable of containing within itself all the energy released during these processes without allowing ionized gases to escape or creating any other external manifestation which could be injurious. The first problem is that of analysing the system behaviour so as to be able to calculate and predict the worst fault conditions which can occur. Even now the analysis of systems and circuitry is not so complete as to make prediction of the worst conditions certain for every case. Fortunately, fuses are easily capable of dealing with situations beyond those which exist in contemporary systems even allowing for uncertainties of calculation.

The capability to interrupt fault currents in a controlled and predictable manner is known as breaking capacity. In addition to possessing this capability the fuse has the additional virtue of being able to operate so quickly as to limit the fault energy to a small proportion of its potential or prospective value. The property of fault energy limitation is almost unique to the fuse and can be achieved in various degrees. Where a particular degree is required and is specified, reliability in this sense implies conformity to the specified limits.

From a design point of view, the passive and active functions are mutually incompatible. For carrying a given current with minimum loss a given type of fuse element needs to be as large as possible in cross section, whereas for fault current interruption, the same type of element needs to be as small in cross section as possible. Although this is an over simplification of the position which ignores the fact that fuse elements are seldom simple conductors, the art of fuse element design is still fundamentally concerned with reconciling these two seemingly incompatible aims. Failure in the passive function may have widely different consequences from failure in the active function, and it is the consequences of failure, rather than the failure itself, which ultimately matters. A complete failure in the passive function is merely an open circuit which may have nuisance value or which may transfer stress to other parts of the system, but which is not often, of itself, of serious consequence. Partial failure during the passive function may be more serious if it leaves the fuse in a weakened condition such that it is unable subsequently to deal successfully with a heavy fault. This risk can be mitigated by design so that partial failures during the passive function, caused by overstressing beyond the legitimate duty, can be made to fail to safety. The best fuse designs do in fact possess this facility, although up to the present time it is not specified in any Standard.

It is particularly important when assessing the reliability of fuses, to clearly differentiate between failure and legitimate operation. It is unfortunate that fuses are often referred to as having failed when they have, in fact, operated or blown in the legitimate duty of interrupting dangerous overcurrents. Failure is properly defined as inability to behave according to claimed performance in respect of one or all of the specified parameters. The inability to interrupt short circuits may have dire consequences. Excessive temperature can cause deterioration leading to unwanted outage, which may be serious in a greater or lesser degree. For those fuses, which are employed specifically to limit fault current energy or transient overvoltage to specified values, failure may be said to have occurred if these values are exceeded. Overvoltage during arcing is another parameter which can be controlled by design and properly specified. If it is inadvertently excessive in service it can overstrain the system insulation and produce a chain reaction of secondary faults.

Fuses must be used within the voltage rating assigned to them and within specified limits of environmental temperature. The conditions of usage are very relevant in deciding the criteria of success or failure and of assessing intrinsic reliability.

14.2.2 Basis of rating

It is important to recognize that the ratings for current, voltage, breaking capacity, peak current, energy limitation and arc voltage are conventionally chosen values relating to closely specified conditions. Where the conditions in service vary in any respect from the conditions specified in the rating tests, correction factors may have to be applied. Usually, the conditions are chosen to be reasonably representative of average working conditions and in most cases carry a sufficient margin of safety to meet the normal contingencies of service. Rationalization on these lines is useful for general practice providing it is recognized that higher degrees of protection are possible for special cases by reference to the fundamental bases of rating.

Current rating is defined as being the maximum current which a fuse will carry continually without exceeding an arbitrarily chosen temperature rise under closely specified environmental conditions. The current rating must also be chosen with respect to the conventional minimum current at which the fuse will begin to melt. There must obviously be sufficient margin between current rating and minimum fusing current to avoid deterioration, although the margin can be varied by design above an acceptable minimum. It must, however, take into account the variations likely to occur due to environmental conditions with regard to any extreme conditions likely to arise.

The fuse is by nature a generator of heat and must dissipate its own watts loss through the surrounding media, either by conduction or convection. The terminals and conductors to which a fuse is connected are also generators of heat and in extreme cases these may produce a higher temperature than the fuse itself, thus exporting heat into the fuse, rather than the reverse. If the fuse is contained in an enclosure with other equipment, all of which is generating heat, the air immediately surrounding the fuse and the general level of temperature in the enclosure will be hotter than the temperature of the outside air. Even where the actual electrical load is light, the frequency and duration of the thermal cycling to which the fuse is subject under these conditions may also be a significant factor in determining the rating of a fuse to avoid deterioration in service.

Voltage rating is particularly important because, in common with all other interrupting devices, the fuse is voltage sensitive during the interrupting period. The arc is the manifestation of the stored energy in the circuit plus the energy derived from the system voltage, the latter component becoming more significant in the later stages of arcing. The successful interruption of the arc may be seen as the ability of the fuse to build up its dielectric strength at a greater rate than the instantaneous system voltage can sustain the arc. For a given set of circumstances, there is a maximum voltage at which a given fuse will safely interrupt. The voltage rating is therefore a value below this, chosen so as to give a safety margin sufficient to cover possible circuit variations, as well as the manufacturing tolerances of the fuse itself.

It is assumed that a fuse can safely operate at any applied voltage below its rated voltage. This is generally true of most contemporary fuses, except for qualifications which need to be made in respect of the arc voltages likely to be produced.

Breaking or rupturing capacity rating presupposes that the fuses will safely interrupt all values of current up to its short-circuit rating and within its rated voltage. It is necessary to impose limiting values of asymmetry in the a.c. circuit and to assume sinusoidal conditions at standard frequency. Satisfactory performance on a.c. is no guarantee of similar performance on d.c. which should always be regarded as being a different category of duty. It is also recognized that the maximum arc energy stresses do not necessarily occur at the highest values of breaking capacity for which a given fuse is rated. There are in many cases, critical currents of a lower order which produce higher stresses. The specification of breaking capacity, therefore, must include a recognition of these possibilities.

The ability of H.R.C. fuses to limit short-circuit currents and energies is well known, but it is only in comparatively recent years that degrees of limitation have come to be demanded for various applications. There is, for instance, good reason for critically limiting short-circuit energies when protecting solid-state devices because in many cases such devices can only be used effectively when protected in this way. Energy limitation in this sense is dependent upon a more critical control of voltage and prospective short-circuit current than is usually necessary for other more ordinary purposes. It does not follow that the highest degrees of limitation are to be desired for every circumstance, because these can only be achieved by compromising other parameters which may in other instances be more important.

When an arc is interrupted an arc voltage is created across it, this being proportional to the rate of change of current and the inductance. Fuses which are designed to chop the arc quickly must produce a relatively higher voltage than those which allow the arc to persist longer. Arc voltage must be limited to values which the system can withstand having regard to its insulation level, and fuses can be designed to limit arc voltage to any desired value consistent with the various areas of application. The criteria of failure and assessment of reliability must involve any or all of these parameters with different degrees of emphasis depending upon the particular application to be considered.

14.2.3 Application

Any evaluation of reliability must assume that the product in question has been applied correctly. If the evaluation is made on the basis of test results, the condition of test must be assumed to relate to correct application. If the evaluation is made on the basis of field statistics, then it must be assumed that all cases of misapplication have been eliminated from the equation. The correct application of fuses involves a knowledge of the circuit and environmental conditions in which the fuse is expected to operate, and the varying circumstances which are likely to occur. Although circuit behaviour can be predicted with reasonable certainty under steady-state conditions, the evaluation of transient conditions which apply during fault occurrences are considerably more complex. The safety margins and rationalizations, which are built into fuse ratings, are intended to allow for errors which are likely to occur in assessing service conditions. In the vast majority of cases, the margins allowed are more than adequate.

A good instance of this principle is that which concerns the problem of ensuring that the fuse has adequate breaking capacity for a particular application. The academic approach to this problem would be to calculate or estimate the fault level of the system and to apply a device having a breaking capacity not less than this value. An alternative approach is to adopt the convention of making all devices of sufficiently high breaking capacity to more than cover any fault level likely to occur and to use such devices in all cases. The latter approach is possible with fuses because they can be designed and manufactured economically to the values of breaking capacity required. This more practical approach is obviously to be preferred because the calculation of system behaviour is time-consuming and usually beyond the capacity of the majority of persons who are responsible for applying fuses.

In observing this principle it is obviously necessary to have regard to parameters other than breaking capacity. Emphasis in various degrees and relating to combinations of parameters are needed in respect of the various types of application now serviced by fuses. For example, the requirements for industrial distribution installations are different from those of public supply networks; High voltage or d.c. systems have to be seen differently from low- or medium-voltage and a.c. systems. Solid-state devices are sensitive in all parameters.

Protection is achieved when the fault energy let through by the fuse is less than the withstand of the equipment to be protected. This presupposes that the withstand of equipment is a known quantity and can be expressed in terms which are directly comparable with the published fuse data. Unfortunately, this is not the case, except for the more sophisticated equipment,. such as solid-state devices, for which such information is vital to the economic operation of the equipment concerned. In other spheres the proof of effective protection has been largely subjective and empirical. The fact that experience has been favourable is largely due again to the safety margins which fuses can provide. Meanwhile, fuse manufacturers have continued to present fuse data in a manner consistent with the practical state of knowledge of the majority of fuse users. This does not mean that the data is necessarily less precise, but the formalization involved, if taken to excess, can lull the user into a false sense of security or otherwise can inhibit the use of fuses in those applications where closer protection is necessary. Fuses are actually much more versatile than is generally supposed. They can be made to high degrees of accuracy and manufacturing tolerances consistent with these can be declared. There is thus no impediment to clearly specify the basis upon which reliability of fuses in any given situation can be assessed. The most difficult part of the problem is that of accurately predicting the duties and conditions of service which a fuse will have to meet, or of assessing, after the event, the conditions which applied during fuse operation.

14.2.4. Type tests

Functional and operational tests on manufactured fuses are obviously not possible as routine because such tests must necessarily be 'destructive'. Performance is therefore assessed on the basis of type or design tests, with the clear implication that fuses offered for sale are in every respect identical with the proven and approved prototypes.

Type tests must necessarily be comprehensive enough to check all parameters of rating and to prove compliance with the relevant specifications. National and international standards exist, which specify test conditions for all parameters in great detail, while at the same time listing the criteria of success and failure appropriate to each.

Many countries insist upon official approval of fuses by independent testing authorities. In some cases this is a statutory requirement, while in others it has become a contractual requirement often imposed by users. In Britain and the Commonwealth, the Association of Short-Circuit Testing Authorities provides such a service and issues A.S.T.A. certificates of rating. Statutory approval procedures are imposed in Canada by the Canadian Standards Association; in the U.S.A. by

Underwriters' Laboratories; in the Netherlands by K.E.M.A.; and by similar authorities in other countries.

The approvals procedures insisted upon by these authorities involve additional rules of interpretation to qualify the standards and specifications. Procedures are also required by which full and precise descriptions of certified prototypes may be recorded so that the identity of fuses offered for sale under the terms of the approval can be checked at any time.

14.2.5 Quality control

Quality control in the manufacture of fuses must be of high order, not only because of the obligation to comply with approvals procedures, but because the failure of a fuse in a high-power circuit can result in disaster of first magnitude. The quality control methods employed need to follow the best practices already established in the manufacture of good-quality electrical hardware, but the standard of inspection and criteria for rejection must be of specially high order in recognition of the fact that the fuse is concerned with providing safety. It may, for instance, be economically admissable to take calculated risks on the failure rate of certain consumer durables, but it is not ethical to admit a possible failure rate with protective devices. Although fuses, in common with all other commercial products, cannot be expected to achieve 100 per cent perfection, the quality control procedures must be based on the assumption that this is possible and inspection must be rigorous enough to detect and reject imperfections where they occur. Thus, the well known edict that quality must be built into the product at all stages from the raw material onwards has rather special significance in the case.

It is possible to institute a number of non-destructive tests on completed fuse-links to check the consistency of manufacture. The measurement of fuse-link resistance is typical of these and can be particularly revealing. If carried out on a 100 per cent basis to predetermined tolerances, it can give a good indication, not only of the accuracy of individual items, but of overall trends. Resistance checking of sub-assemblies at various stages in manufacture is an economic necessity for some of the more expensive fuses, but also contributes to the general standard of quality.

The inspection of completed fuses by radiography is another method used extensively. For some fuses, such as high-voltage or aircraft fuses, a 100 per cent inspection may be necessary because of the vital nature of the application. For other fuses a smaller percentage sampling can be justified, although it is usual and prudent to subject pilot manufacturing runs to 100 per cent inspection before progressively reducing the inspection rate.

The statistical methods of analysing rejects, which are normally used as a tool for controlling the economics of manufacture also have extra significance where fuses are concerned. They can and should be seen in relation to standards of safety which the fuse is expected to provide. This means that the distinction between acceptance and rejection must be clearly defined so as to avoid the possibility of 'grey areas', which leave room for discretion by those who may be more concerned with output than the safety of the ultimate user.

'Destructive' testing of finished samples is conclusive proof of quality and this includes testing for breaking capacity and its related parameters under conditions of full power. Such tests can be expensive because of the size of the power source which is needed, but up to the present time there seems to be no reliable method by which the test conditions can be simulated. The provision of a high-power testing station capable of delivering short-circuit power of up to hundreds of megawatts is essential to cover contemporary needs.

14.3 Evaluation of reliability of 'English Electric' fuses

The need for reliability evaluation has always been recognized in a qualitative sense and the task now arising is to quantify the information available into terms which can be processed according to the newer statistical techniques. The first problem is to establish the statistical basis and to ensure that this is completely rational and meaningful in relation to the purposes for which fuses are used.

The purpose of this account is to collate existing information under a number of arbitrary headings in an attempt to make a first approximation of reliability in statistical terms.

This is intended only as an introduction to the subject. Considerably more study and research is required before valid data relating to fuse reliability can be made available.

14.3.1 Quality control and inspection

Although the English Electric Co. produces fuse-links at the rate of over ten million per annum, the variations in type and rating are such that all methods of manufacture from mechanical mass production to small batch production are necessary. The size and variety of the operation makes it impossible to give more than a general idea of the measures adopted towards quality and reliability, but it may be accepted that the most up-to-date concepts are constantly sought after. The most compelling proof of success is the high reputation and commercial success enjoyed by the Fusegear Division over the last thirty years. This provides incentive for improvement and imposes responsibility of stewardship.

Quality control is a recognized function in its own right. Design and manufacturing standards are closely specified. Weekly analysis of 'rejects' or variations from optimum limits are made under a variety of agreed headings. Histogram charts are kept to show the trends and for general reference.

Formal meetings by a 'quality-control' action group representing design, manufacturing process, production, quality control and inspection functions meet at regular intervals.

The immediate purpose of these activities is to maintain the efficiency of manufacture. The standards of quality stem from initial design concepts and the manner in which these are specified, but constant feedback from quality-control is an essential element in improving standards and controlling those areas where it is necessary to allow discretion in implementing instructions.

244

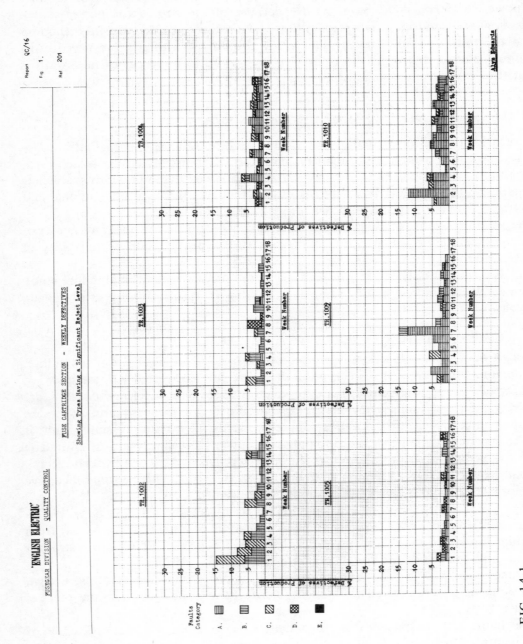

FIG. 14.1
Typical histogram chart

The sampling of fuses and components is a constant activity. For the most part this is random, both as regards time and method of selection. Where outside approvals authorities are involved they lay down their own requirements. Large quantities of fuses are regularly monitored by Canadian Standards Association, Underwriters' Laboratories, U.S.A., and for contractual purposes by the G.P.O. Defense Departments and other public bodies.

All field complaints are comprehensively investigated and fed back where appropriate for quality-control action.

14.3.2 Field experience

Customers' queries and complaints and other field information is recorded and analysed on a regular basis. All cases are comprehensively investigated, irrespective of their relative economic importance and this can be justified because the total number of cases to be dealt with is small in relation to the total output in an equivalent period. The information which these investigations produce is valuable feedback for quality evaluation purposes and becomes more useful with the passage of time. Records have existed for very many years and frequently from the inception of the product. See Fig. 14.2.

Many of the queries investigated are those in which users require advice on the reasons why fuses have operated in particular cases. This service is offered to users primarily to promote the use of fuses but provides unique opportunities for obtaining information on fuse behaviour in service. The reasons why fuses have blown are not always obvious at site but an analysis of the blown fuse in the laboratory can often reveal the conditions in the circuit obtaining at the time of the incident. The information thus obtained enables the user to correct his system to obtain better co-ordination and to avoid unwanted outages. See Fig. 14.3.

Tables 14.1 to 14.4 give analyses under a variety of categories of records kept over a ten-year period ending early in 1968. It can be seen that relatively few of the incidents reported can be categorized as fuse failures and the rate at which these occur has remained sensibly constant over the decade considered. The types introduced prior to or at the beginning of this period showed no tendency to fail more frequently by the end of it. This supports the contention that there is no significant deterioration effect in well-designed, well-applied H.R.C. fuses which would tend to limit their service life. New designs of fuse tend to have an initially higher incidence of reported queries. This is almost certainly due to the teething trouble of application and as the snags are sorted out the reported incidents revert to the random pattern of failures typical of established designs.

Generally speaking, those types showing a significantly high failure rate are the larger or more costly fuses, or alternatively those which are used for the more important applications. It is probable that the higher rate reflects the greater likelihood that faults in these cases will be more readily reported than with other types. This leads to the conclusion that for the other types there is a proportion of

FIG. 14.2

Typical monthly summary of field investigations

Customer	Parts affected	Nature of query	Cause	Action	Remedy	Remarks
A	Fuse-links type TIA30 (two)	Claimed to be open-circuit when supplied	Blown on motor-starting; heavier rating required	Report with radiograph sent to local tech. rep.	Fit larger rating	No failure
B	Fuse-links type K4PCX150 (two)	Request for assessment of fault current	Operated due to short-circuit current	Report with radiograph sent to local tech. rep.	—	No failure
C	Fuse-links type GS 150/300 (two)	Request for examination for possible deterioration	Fuse-links intact and complied with test figures; no evidence of deterioration	Report with radiograph sent to local tech. rep.	—	This installation has a history of unexplained fuse operation
D	Fuselinks type TLM 750	Request for assessment of operating conditions	Fuse-links operated correctly due to heavy overload	Report with radiograph sent to customer	—	Advisory service; no failure
E	Fuse-links type TIA10 (eight)	Operated for no reason when in circuit with 0.5 hp motor (415V 3-phase)	Radiograph revealed operation at 17/25 A; probably due to flashover on system	Report with radiograph sent to customer	—	No failure
F	Fuse-fittings type RS400-H	New fittings over-heating at 100 A load	Steel pressure plates wrongly fitted under fuse-link tags by customer	Tech. visit to site and corrective action taken	—	No failure
G	Fuse-link type TTM500 (one)	Failed to operate on fault, back-up circuit-breaker tripping out instead	Fuse-link intact and satisfactory to test figs; circuit data obtained which revealed widely incorrect settings of the circuit-breaker	Report with curves and comments sent to customer Intact fuse-link returned	Revision of circuit settings	No failure
H	Fuse-links type D51 (1 A) (12)	Said to operate for no apparent reason	Only one fuse-link blown, other complying with test figs; blown sample interrupted high fault current; examination of circuit diagram revealed the necessity for modification	Report with radiograph sent to customer	Change of circuit or use of heavier fuse rating	No failure

	Item	Complaint	Finding	Report	Action	Conclusion
I	Fuse-link type S.1485 (one)	Unusually high voltage drop across fuse-link	Inadequately secured internal connection	Report with replacement sent to customer	Alert inspection carry out check tests	Primary failure. Production carefully checked but no fault found; concluded to be an isolated instance
J	Fuse fittings type RS.100H	Overheated in service	Overheating due to cable having been damaged in vicinity of base contact	Report sent to customer	Renew cable, and prevent abrasion	No failure
K	Fuse-links type SRC5002 (eight)	Claimed to fail after a short period in use	One intact and complying with test figs; other seven blown at various fault levels; also statement that Normal Load exceeds current ratings of fuse-links	Report with radiograph sent to customer	Use larger rating	No failure
L	Fuse-links type JS500 (no sample)	Did not operate in line with published data	Investigation reveals that customer has compared AC5 fuse-links with AC4 curves	Report with new curves sent to customer	—	No failure Customer reports that test results line up with new curves
M	Fuse-link type GSG1000/45 (One)	Faulty when supplied	Operated due to short circuit	Radiograph and report sent to customer	—	No failure
N	Fuse-link type 197.TS.50/50 (two)	Request for assessment; used for special purpose in ignitron circuit	Operated correctly due to overload. Circuit details examined and fuse-links found to be too high in rating	Report sent to local tech. rep.	Replace by 35 A rating	No failure

continued overleaf

Ref	Fuse-link	Complaint	Investigation	Action	Recommendation	Conclusion
O	Fuse-links JP400 TKF300; TF200	Failed to provide proper discrimination	Characteristics of the fuse-links in question are not compatible for discrimination other than at low values of fault current	Report sent to local tech. rep. with suggestion for a circuit revision	—	No failure
P	Various 'T'-type fuse-links used in 440 V d.c. circuits	Failures experienced particularly with TIA and TIS fuse-links of 30 and 35A rating when used to protect highly inductive circuits	Fuse too small in rating and fuse operation occurred due to starting currents and when full inductance was in circuit; overload devices found to be most unreliable	Tech. visit to site; revision of fuse ratings advised, with the point that Type THA and THS fuse-links may be required in the event of further instances of failure due to heavy circuit inductance	Use of correct fuse ratings and improvements of overload devices on equipment	Failures due to incorrect application
Q	Fuse-links type: TLM800 (three) (two intact, one broken)	Said to have exhibited body cracking when carrying normal circuit current	Investigation revealed that 'normal current' was in excess of fuse-links rating. Fuse-link operated at approx. 1000 A and cracking was due to thermal stress, no danger being caused	Report with radiograph of blown fuse-link sent to local tech. rep.; two intact fuse-links returned	—	Customer has now revised his circuit, to remove overload hazard; no failure
R	Fuse-links type TF200 (two)	Request for assessment of operating conditions	One sample operated due to short circuit and the other due to overload	Report with radiograph sent to customer	—	Advisory Service; no failure
S	Fuse-link type TIA30 (one)	Indicator failed to operate	Radiograph revealed that the fuse-link had interrupted an overload correctly; the action of the indicator had been impaired by absorbing moisture	Report with radiograph sent to customer	Prevent storage or use in excessively damp surroundings	Second-degree failure, re-conditions outside works control
T	Fuse-link type TF160	Request for examination of fuse-links and to advise whether this rating is suitable for use with a 75 hp motor	Examination shows that fuse-links have operated under motor-starting conditions, and that a 200 A rating of fuse-link is more suitable	Report with radiograph and recommendations sent to local tech. rep.	Fit 200 A rating	No failure

QUERY RECORD SHEET

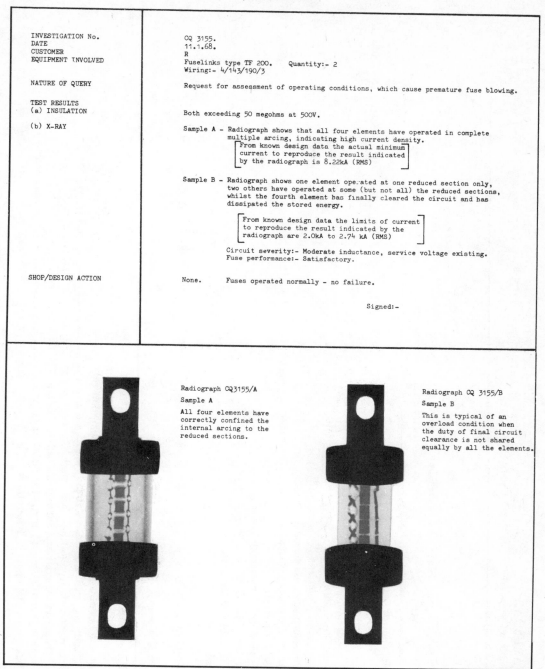

INVESTIGATION No.	CQ 3155.
DATE	11.1.68.
CUSTOMER	R
EQUIPMENT INVOLVED	Fuselinks type TF 200. Quantity:- 2
	Wiring:- 4/143/190/3
NATURE OF QUERY	Request for assessment of operating conditions, which cause premature fuse blowing.
TEST RESULTS	
(a) INSULATION	Both exceeding 50 megohms at 500V.
(b) X-RAY	Sample A - Radiograph shows that all four elements have operated in complete multiple arcing, indicating high current density.

Sample A - Radiograph shows that all four elements have operated in complete
multiple arcing, indicating high current density.

> From known design data the actual minimum
> current to reproduce the result indicated
> by the radiograph is 8.22kA (RMS)

Sample B - Radiograph shows one element operated at one reduced section only,
two others have operated at some (but not all) the reduced sections,
whilst the fourth element has finally cleared the circuit and has
dissipated the stored energy.

> From known design data the limits of current
> to reproduce the result indicated by the
> radiograph are 2.0kA to 2.74 kA (RMS)

Circuit severity:- Moderate inductance, service voltage existing.
Fuse performance:- Satisfactory.

SHOP/DESIGN ACTION   None.   Fuses operated normally - no failure.

Signed:-

Radiograph CQ3155/A

Sample A

All four elements have
correctly confined the
internal arcing to the
reduced sections.

Radiograph CQ 3155/B

Sample B

This is typical of an
overload condition when
the duty of final circuit
clearance is not shared
equally by all the elements.

FIG. 14.3

Typical customers' query record sheet (ref. case R in Fig 14.2)

TABLE 14.1
Analysis of reported incidents by fuse type

Type of fuse	Output × 10⁶	Total reported complaints	Total justified complaints	Justified complaints reported per 10⁶ fuses	Estimated total justified complaints per 10⁶ fuses
Medium voltage industrial for U.K.	35.0	330	70	2.0	10.0
Medium voltage industrial small ratings	25.5	95	45	1.8	9.0
Supply distribution	3.0	46	6	2.0	10.0
Sub-circuit protection	5.0	22	5	1.0	10.0
Domestic and house cut-out	5.0	5	4	0.8	8.0
Aircraft, traction, marine and interservice	9.0	65	15	1.7	5.1
Fuses to continental and N. American specification	5.5	16	6	1.1	5.5
Fuses for semiconductor protection	2.5	52	15	6.0	12.0
High-voltage types	0.6	95	8	13.5	27.0
Trip fuses	0.5	23	11	22.0	66.0
Total or average	91.6	749	185	2.0	9.25

Period covered: 10 years, 1958–68.

Note: Difference between 'reported' and 'estimated' complaints columns is due to the application of a correction factor to allow for the assumption that only a proportion of fuse incidents are actually reported. For reasons already stated it is assumed that incidents affecting high-voltage and semiconductor applications are more frequently reported than for industrial low-voltage applications. Aircraft and interservice fall in between. The table assumes a ratio of 2:1 between 'reported' and 'estimated total' for high-voltage and semiconductor types: 3:1 for aircraft and interservice and up to 10:1 for others.

TABLE 14.2

Analysis of justified reported incidents by type of fault

Category of failure	Nature of failure	No. reported*
Primary functional	Fuse did not operate to clear fault condition	1
Primary functional	Fuse shattered while clearing fault	24
Primary functional	Hole burnt in end-cap of fuse during operation	12
Primary functional	Flashover during operation	6
Primary functional	Low insulation resistance after operation	2
Primary functional	High resistance or excessive temperature rise on normal load	34
Primary functional	Loose or detached tags or blades	34
Secondary functional	Maloperation of trip fuse or striker	32
Secondary functional	Indicator wire remained intact after fuse operated	5
Secondary functional	Maloperation of indicator components	12
Non-functional	Faulty labels, markings, etc.	23
Total		185

Analysis covers 10-year period 1958–68, during which total output of fuses was over 91 million.

*See Table 14.1 for note on probable relationship between 'reported' and 'estimated actual'.

TABLE 14.3

Analysis by type of (justified) reported failure for aircraft and interservice fuse types only

Category of failure	Nature of failure	No. reported
Functional	Fuse did not operate to clear fault condition	none
Functional	High resistance or excessive temperature rise on normal load	3
Functional	Loose or detached tags	3
Non-functional	Faulty markings, etc.	2
Total		8

Period covered: 10 years.

Output of these types during this period: 7 million.

From above, functional failures per 10^6 fuses of this type: 1.15 (from reported faults only). See notes to Table 14.1.

Note: No case was reported of 'failure to danger'.

TABLE 14.4
Analysis by functional category

Function	No. of justified reported failures	Failure per 10^6 fuses
Passive function failure	68	0.74
Active function failure	45	0.49
Secondary function failure	49	1.0*
Non-functional failure	23	0.25
Total functional failures	162	1.76
Total primary functional failures	113	1.24

Period covered: 10 years, 1958–68.
  *A large proportion of fuses do not have strikers or other indicating means. Hence 'Failures per 10^6 fuses' relates only to that proportion of the total fuse output which includes such features.

failures which are not reported. Some adjustment in the figures is therefore necessary to account for the discrepancy.

On the other hand, all faults are not of the same importance. Hence the analysis differentiates between those faults which are of primary and those which are of secondary importance. These being designated 'primary functional', 'secondary functional' or 'non-functional'. A functional fault is one which may impair the ability of the fuse to perform its functions (as distinct from non-functional faults such as detached label, incorrect marking, etc.). A primary function fault is one which affects either the passive (current-carrying) or active (fault-current-interrupting) roles. Secondary functions are held to be those in which the operation of ancillary features such as the indicator wire in a medium-voltage fuse or the striker of an high-voltage fuse are concerned.

14.3.3 Analysis of test evidence

Field experience shows that heavy short circuits are comparatively rare. Most fuses actually blow on moderate faults. This fact does not make the H.R.C. fuse any less necessary because it is impossible to predict when the heavy faults may occur on a system where they are potentially possible. It is known that a very large proportion of the fuses manufactured each year are used for replacing those which have cleared faults of one sort or another and that of these a considerable number must have experienced heavy duty, but it is difficult to estimate how many fuses actually perform the ultimate duty for which they are designed. A more reliable estimate of performance in this region can be found in an analysis of the sampling and other tests which have been carried out continually over a long period of years.

Some 20 000 high-power, short-circuit tests of H.R.C. fuses have been carried out by the English Electric Co. over the last ten years or so. While most of these were

TABLE 14.5
Estimated failure rates from test evidence

Fuse group	Total no. of samples tested	Passed	*Failed
Medium voltage, industrial, public supply, domestic, etc.	1019	1012	3
High voltage	313	311	2
Types for semiconductor protection	193	191	2
Medium-voltage types tested in other manufacturers equipment (e.g. with contactors, etc.)	5000 (approx.)	4997 (approx.)	3
TOTAL:	6525	6511	10

Failure percentage: 0.15%.
Period covered: 10 years, 1958—68.
*These figures are necessarily rough approximations because of the difficulty of establishing the cause of initial failure where explosive damage has resulted and evidence destroyed.

prototype or development tests, about 10 per cent were repeat tests carried out on stock fuses, previously certified. In addition, some 5000 tests have been carried out and recorded by other manufacturers who incorporated 'English Electric' fuses in their equipment. The results show that approximately 0.15 per cent of all these 'retests' were unsuccessful, although failure in some of these incidents was due to the test conditions having been open to dispute. In all cases the test conditions and the criteria for success were much more stringent than service duty and failure on test does not necessarily mean that failure would have occurred in service. All test failures, whether disputed or not, have resulted in design change or the elimination of doubtful components or materials. Where doubt still existed, the certificates of short-circuit rating have been rescinded.

Table 14.5 summarizes the results of the tests referred to so far as these can be estimated.

14.3.4 Case investigations

A number of representative installations have been specifically surveyed for the purpose of obtaining statistical information. These investigations supplement and substantiate the existing field and test evidence.

The following typical case illustrates the trend:

Installation: Engineering plant set up for mass production of consumer durables on continuous flowline system. Electrical distribution from local

TABLE 14.6
Survey of H.R.C. Fuse Usage In Industrial Plant

Question	Answer
No. of H.R.C.fuses installed	6500
No. of replacements per annum	610
Cases of known fuse maloperation	none
Length of experience in plant concerned	10 years
Mean life of H.R.C. fuse	10.5 years

substation at 415 and 240 V a.c. Electrical load comprises variety of medium and heavy machine shop equipment, automatic processing equipment, control gear, heating and lighting services, etc., etc.

The information was collected by directly questioning maintenance personnel whose experience extended over a sufficient time to be relevant. The job of the questionnaire was to obtain an estimate of the mean service life of H.R.C. fuses in general and to seek information concerning fuse behaviour both favourable and unfavourable.

A closer analysis of the replies received reveals that the percentage replacement per annum was much higher for small ratings used in final sub-circuits (e.g. plugs, etc.) than for the larger ratings. Small fuses used to protect hand tools and small machines are called upon to deal with a much greater frequency of faults and, because of their relative cheapness, are also often used as diagnostic tools to save time in locating faults. This is malpractice but involves no danger because of the quick action of the H.R.C. fuse. In some cases several fuses may have blown on the same fault before it is cleared. The 'mean fuse life' is different for sub- and sub-circuit fuses than for main circuit and busbar zone fuses, the former being a mean life of something less than five years, the latter considerably exceeding five years.

It may be noted that the plant electricians questioned were unable to recollect a single instance of intrinsic fuse maloperation over an aggregate of several decades of experience. Disastrous incidents which stood out in the recollections of the electricians had, in all cases, been traced to causes other than primary fuse failure. A number of cases were reported of indicator devices failing under damp conditions and other cases have occurred due to excessive contamination.

Another case study has also been made of a large nuclear research establishment. This survey is different in form from the previous case described in as much as the statistical information is available from systematic records. A system of fault reporting, recording and analysis has been operated since normal running commenced some nine years ago.

The research establishment and its reactors form part of a completely independent unit, again fed from a local 11 kV/415 V substation. The installation includes

TABLE 14.7

Analysis of Fuse 'Incidents' at Nuclear Research Establishment

Incident	No. of cases in 9 years
Fuse operated for reasons not recorded in fault report:	
Standard industrial	108
Small types for protecting electronic equipment	32
Types for solid-state protection	none
Total all types	140
Broken tags or loose connections (all types)	4
Other faults	none
Total faults	144

Period covered: 9 years, 1959–68.

all the services for operating a number of reactors with metering control alarm and other ancillary circuits, together with the usual plant services associated with a large establishment. H R.C. fuse-links are widely used in all the circuits referred to, the quantities being as follows:

Standard industrial fuses for motor and distribution circuit protection	1500
Small fuses for protecting, metering and electronic gear etc.	1000
Special fuses for protection of solid state equipment	100
Total	2600

Table 14.7 gives the results of an examination of fault report records over nine years.

No incident has been reported of a fuse failing to danger in a way which could cause consequential damage. The majority of incidents are cases of unexplained fuse operation in which it is evident that the fuse has really blown in doing its legitimate duty. The replacement of fuses in these cases has been regarded as routine, particularly where the reason for the fuse operation is obvious. The number of fuse replacements has been fairly constant over the nine-year period considered, which is further evidence that deterioration is not a significant factor. The items headed 'loose connections' cover a variety of malfunctions, mainly due to bad connections rather than to any deficiency in the fuse-link itself.

14.3.5 Calculation of failure rate of H.R.C. fuses

The data available from various sources provides a fair basis for calculating to a first approximation the intrinsic reliability of H.R.C. fuses as circuit components.

Failure rate in terms of number of faults per component per million hours is given by

$$\frac{f \times 10^6}{nt}$$

where f is the number of failures, n is the total number of components (fuses) and t is the period of test or service life.

It has been shown that the service life of fuses varies considerably in practice according to application. The passive, or current-carrying, life on the other hand is indefinite, as has been proved by investigations carried out on fuses which have been in continuous use for more than twenty-five years. From Table 14.6 a mean fuse service life of 10.5 years or 92 000 hours has been provisionally adopted for the calculations which follow.

Method 1 Failure rate calculated from analysis of field data. From Table 14.1, estimated failures are almost five times the number of complaints actually received; it is right to err on the pessimistic side. Failure rate is then:

$$\frac{9.25}{10^6} \times \frac{10^6}{92\ 000} = 0.0001 \text{ failures per } 10^6 \text{ hours.}$$

Method 2 Failure rate calculated from test evidence. From Table 14.5 estimated failures are 10 out of 6525 fuses tested. Assuming each fuse had completed its average service life span of 92 000 hours before test, failure rate is given by:

$$\frac{10}{6525} \times \frac{10^6}{92\ 000} = 0.016 \text{ failures per } 10^6 \text{ hours.}$$

Method 3 Failure rate calculated from industrial plant survey. From Table 14.6, no primary fuse failures were reported out of 6500 installed.

Adopting the statistical device of assuming that the 6501st fuse would have failed, a failure rate can be derived:

$$\text{failure rate} = \frac{1}{6501} \times \frac{10^6}{92\ 000} = 0.0017 \text{ failures per } 10^6 \text{ hours.}$$

Method 4 Failure rate calculated from survey at a nuclear research station. From Table 14.7, no primary fuse failures were reported out of 2600 installed. Again assuming that the 2601st fuse would have failed:

$$\text{failure rate} = \frac{1}{2601} \times \frac{10^6}{92\ 000} = 0.0042 \text{ failures per } 10^6 \text{ hours.}$$

14.3.6 Conclusion

It may be inferred that had the samples been larger in methods 3 and 4 the failure rates would have tended towards the lower value given by method 1. Hence, there is probably sufficient justification for stating at this stage of the investigations that the failure rate of the H.R.C. fuse, when used under normal service conditions lies somewhere between 0.001 and 0.0001 faults per million hours. This compares favourably with published figures for other items of equipment; e.g. Failure Rate Tables (U.K.A.E.A.) quote 0.1 for wire connections, 0.02 for soldered joints, 3.0 for contactors and 2.0 for circuit-breakers (all as before in faults per million hours).

15 H.R.C. Fuses of earth fault protection of medium-voltage industrial systems

Whether the H.R.C fuse will provide earth fault protection depends upon the state and condition of the return earth path between the fault and the source of supply.

Where solid earthing is the norm, fuses can provide good earth fault protection provided the earth path is adequate and kept so.

The wide variation of medium-voltage industrial systems, some having solid earthing and others not, makes it difficult to draw up regulations and codes of practice to cover all situations throughout the world but there is a general desire to achieve international harmonization on these issues. The I.E.E. took the initiative in 1972 to call a conference at which the advocates of various systems could be heard. It may be many years before final agreements can be reached and rationalization and harmonization effected. Meanwhile it is necessary to reiterate the excellent service record achieved in British systems in which the H.R.C. fuse has played a prominent role.

Paper presented to the I.E.E. Conference on International Medium Voltage Earthing Practices in London, March 1972 and originally published by the Institution of Electrical Engineers.

15.1 Introduction

The problems which arise in providing earth fault protection are not new but re-examination of established practices is necessary from time to time because of changing circumstances.

At the present time attention is being focused in international circles upon the need to safeguard domestic users of electricity against the hazards of electric shock. Presumably experience in certain countries has not been so favourable as in the U.K. The controversy which has arisen from this situation has coincided with a desire within I.E.C. to produce international wiring regulations and codes of practice. Their problem is to produce regulations which are simple enough to be applied by uninformed people, while at the same time being comprehensive enough to prevent danger in all circumstances. As a result existing national regulations are being brought into question and earth fault protection in particular is under scrutiny. This is a subject which is not amenable to simple rules and misapprehensions already exist.

Protection against shock hazard requires a different approach to that needed for the prevention of thermal and other damage. It is also necessary to distinguish between those situations in which good earth conductivity exist or can be made to exist and those where it does not. The difference between 'earth fault' and 'earth leakage' needs to be defined. Other variables may apply due to site conditions, human factors and scale as for instance in domestic or industrial situations or in supply networks. In other applications, such as marinecraft, aircraft or explosive situations, the requirements may differ widely both in the manner of protection and degree.

Although incentives now exist for a re-examination of the methods by which earth fault protection can be achieved, and although regulations need to be brought up to date, the fact is that many existing practices are fully effective and are likely to remain so without change. Earth fault protection by H.R.C. fuses is a good example of this. The success of H.R.C. fuses in this role is based upon the simple fact that the majority of industrial installations in the U.K. and in other areas of the world where British practice is followed depend upon them exclusively for earth fault protection and have done so for many years with entirely satisfactory results.

15.2 Factors affecting choice

There is no single method or universal panacea which will provide complete earth fault protection in all cases. Fuses are successful in industrial circuits because the earth fault currents are substantial and solid earthing is the rule rather than the exception. Fuses cannot provide complete immunity from shock hazard where the site conditions are unfavourable to their use although this is a circumstance which occurs rarely in practice. In cases where the earth path is doubtful E.L.C.B.'s may provide the only solution. 'Earth leakage' circuit-breakers are aptly named for their function and the choice of types is wide to provide varying degrees of sensitivity for different applications. For portable appliances double insulation may be the best practical solution and the use of low voltage may be the best method in the case of portable tools.

Industrial installations, supply networks and domestic circuits each pose their own problems, while special consideration has to be given to applications such as those involving ships, aircraft and situations in which explosive atmospheres occur. There are also differences in the degrees of hazard present under each of these headings and the degrees of acceptable risk may also be a variable factor.

The success of any fuse in providing earth fault protection depends upon the presence of a return earth path of sufficiently high conductivity to permit a fault current large enough to blow the fuse in reasonable time. The present I.E.E. Wiring Regulations reflect this but because they have to be written to embrace all types of fuse in use they do not recognize the fact that the H.R.C. variety is several orders of magnitude more sensitive than the simple open wire type of fuse. The fact that the H.R.C. fuse provides very much better earth fault protection than the regulations recognize explains their widespread use in industry and the satisfactory results which have been obtained over several decades.

The protagonists of earth fault protection fall roughly into two groups, those who advocate the solid earthing of systems and those who regard earth as merely a datum of potential. The latter place no reliance upon it to conduct substantial earth fault current. The advocates of solid earthing are greatly in the majority in this country and in many other parts of the world. It seems to be generally agreed among such advocates that where solid earthing is adopted it should be positive and adequate, otherwise it is better to treat the system as if earth conductivity was of low order. Any attempt to try to steer a middle course is likely to introduce more hazard than it is intended to prevent. Protective systems which depend upon the integrity of the earth path do so throughout the life of the system and this implies that the integrity must be beyond question. In recent years protective multiple earthing (P.M.E.) has been gaining ground particularly for supply networks. This system has already been proved to be economical and reliable and has already made a very significant contribution to the safety of systems.

15.3 The fuse

Where solid earthing is established the H.R.C. fuse gives earth fault protection to a high degree whatever the magnitude of the fault. The well-known ability of the fuse to limit fault energy is as relevant for earth faults as for phase to phase faults. In some cases it is more so, because the earth path can be somewhat more vulnerable than the primary phase conductor. Time/current characteristics can be designed into the fuse to suit a variety of circumstances. Once the fuses have been manufactured to produce particular characteristics, they will remain faithful to them throughout their service. Since the performance is predictable to such a high degree it is easily possible to calculate or assess the degree of protection which can be provided. Simplicity, fidelity to characteristics, non-deterioration and overall reliability are the key notes.

H.R.C. fuses properly chosen can easily limit fault stresses to values well within the withstand of the equipment in the system, thus there is in practice very little difficulty in providing complete protection against thermal or mechanical damage. The criticisms levelled at fuses are invariably those which relate to the elimination of shock hazard during the time it takes the fuse to interrupt the circuit. During this time differences in potential can arise between metal parts which become alive during an earth fault and the surrounding earth. Where the earth fault currents are low, the operating time for the fuse can be relatively long but the potentials relatively small. If the earth fault currents are high, then the potentials can also be high but the operating time for the fuse is correspondingly short.

There are very few recorded cases of danger to personnel arising from this cause, despite the fact that millions of circuits in use have no protection other than H.R.C. fuses. The danger appears, therefore, to be more theoretical than actual. The total number of earth faults which occur compared to the number of circuits in use is relatively small presumably because of the high standard of installation practice. The number of these where personnel are directly at hazard must, by conjecture, be smaller still. The majority of earth faults are of sufficient magnitude to blow fuses very quickly because of relatively good earthing practice. In a large proportion of cases the earth fault escalates into phase fault particularly where three-phase cables are involved and this increases the magnitude of the fault current very rapidly to cause almost instantaneous fuse operation.

The situation is further helped by the fact that the earth environment adjacent to fault is sufficiently equipotential in most situations.

15.4 Other methods and devices

There are several methods available other than fuses of providing earth fault protection. In general they come under the headings of E.L.C.B.'s, double insulation and the use of low voltage. These methods are not necessarily alternative to fuses because any or all of them may be used in combination on the same system to cater for the needs of circuits in particular locations.

Where the earth path is doubtful or where the only return path available is the earth itself, some form of earth leakage circuit-breaker is usually the only method available which will give satisfactory results. The choice of E.L.C.B.'s is wide and large numbers of these devices are employed in parts of the world where conditions are favourable to their use. Current-operated devices find more favour than voltage-operated devices except for special applications, but there is some controversy concerning the degree of sensitivity which is considered to provide essential safety particularly in domestic premises. Too high a sensitivity results in nuisance tripping and too low a sensitivity prolongs the time during which shock hazard may exist. The breaking capacity of E.L.C.B.'s and their ability to discriminate is limited because of their greater sensitivity. A lack of discrimination between main circuits and sub-circuits of an installation may introduce more dangers than the temporary hazards introduced by earth faults. In domestic circuits lights may be affected if a fault occurs on a power circuit or vice versa. In industry lack of discrimination cannot be tolerated because of the greater dangers associated with moving machinery and also because of possible disruptions to production, etc.

Because E.L.C.B.'s incorporate moving parts they require maintenance and calibration throughout their lives. They are more expensive than fuses in first cost and the additional maintenance adds further cost but is secondary to the danger which may arise where maintenance is overlooked or neglected.

Double insulation and the use of low voltage are necessary for particular applications and although very important in this role cannot be seriously contemplated as applying to the system as a whole.

A great deal can be done in any situation in creating environmental conditions to prevent shock hazards. This applies particularly in industrial situations and some study towards this end is overdue. In most cases the metalwork in the vicinity of electrical machinery is solidly bonded but the practice is not invariable. The creation of an equipotential zone does not of itself eliminate danger but it can substantially minimize it. It is necessary, for instance, to consider the possibility of potentials arising between the boundary of the bonded zone and the surroundings beyond it. The alternative of providing an insulated environment is also worthy of study. This can often be in addition to the equipotential environment and does not necessarily conflict with it.

15.5 Earth path

In industrial practice the earth path is invariably a direct metallic conductor between the protected zone and the sources of supply. Full use is made of metallic sheathing and armouring of cables but there is an increasing preference for the provision of a specific earth conductor. The system is connected to 'mother earth' through earth plates and rods to minimize shock hazard. It is not expected that earth currents of any magnitude would flow through the earth itself. In practice there is very little difficulty in providing a solid metallic earth path with sufficient conductivity to satisfy the needs of protection. The cross-sectional area of the earth path conductors occur in direct proportion to the cross-sectional area of the live conductors. In heavy current circuits the earth path impedance needs to be in the order of milli- or microhms and this offers no economic problem. The most critical parts of any system are the joints and connections and in this respect the earth path is no exception. Joints and connections must be capable of withstanding the very considerable short-circuit stresses which arise from high power faults. The stresses are proportional to the square of the current and it is in this context that the value of a current-limiting device such as a fuse becomes obvious. The limiting of stress can often be turned to significant economic advantage in the design and provision of earth path joints and other components. The considerable savings which result are achieved without any compromise on safety because the reduction in severity occurs by several orders of magnitude and even where the cost is halved the safety margins can still be more than doubled when compared to a situation in which the protective device permits the full prospective fault current to be realized.

The testing of earth paths at the time of installation and perodically thereafter is important. Measuring impedance is relatively simple by one of the many methods of loop testing, etc., which have become established and accepted. Testing the earth path as regards its short-circuit capability is impracticable because this can only be done at full prospective short-circuit power. Representative type testing of conductors, joints and other components is the only way to determine short-circuit capability. Where non-current limiting protective devices such as circuit-breakers are used on high-voltage systems it is not unusual to impose impedance into the earth path for the purpose of reducing short-circuit stresses, but this is impracticable on low- and medium-voltage systems because of the possibility of introducing undesirable effects and particularly of increasing fire hazard. It is in this situation that the H.R.C. fuse has particular value. Provided the short-circuit capability of earth path components can be determined to a reasonable degree. protection is assured because the let-through energy of the fuse is usually of such a low value as to permit more than adequate safety margins. Many installations in the past have been commissioned without particular regard to the short-circuit capability of the earth path, but in spite of this the service record has been good because of the presence of H.R.C. fuses in the system presumably chosen in the first place for entirely different purposes.

One group of earth path components which does not always receive the attention it deserves is that which includes the enclosures of switchboards and other fabricated structures. In some cases an earth conductor is designed into the system but in others the structure itself is relied upon to provide adequate earth continuity. Whether an earth conductor is present or not, it is desirable that all parts of the structure are properly bonded together and it is essential to test and measure the impedance of such structures in this regard. The following examples refer to two types of production switchboard and the impedance values are representative of those which have been obtained in practice over more than twenty years of production. Every switchboard is tested as routine and impedance measurements are recorded between the incoming cable gland and each outgoing gland. For proper fuse operation the limit of impedance in any path should be 0.01 Ω. The figures show that there is no practical difficulty in achieving values of much lower order.

15.6 Routine earth continuity checks

Ammeter/voltmeter method at 10 A constant current measured between incoming gland and each outgoing gland.

(1). 800 A 'Combination' Fuse switchboard: standard production		(2). 1200 A 'Frontier' Fuse switchboard: standard production	
Outgoing glands	Resistance (Ω)	Outgoing glands	Resistance (Ω)
A	0.0018	A	0.0015
B	0.0024	B	0.0011
C	0.0023	C	0.0010
D	0.0018	D	0.0011
E	0.0017	E	0.0012
F	0.0019	F	0.0012
		G	0.0010

The manner in which earth continuity is achieved in such fabricated structures rests with the designer. In the example shown the components of the structures are painted before assembly. It is thus necessary to clear or penetrate the paint to achieve electrical connection. This is done by the use of substantial serrated washers under each bolt-head and nut clamping the parts together. The washers not only pierce the paint, but penetrate the surface of the steel thus providing hundreds of point connections in any one switchboard. The washers are substantial enough to permit dismantling and reassembly without impairing their effectiveness to cater for any contingencies during installation or subsequently. These requirements are rarely stressed in specifications for equipment. Tests should be insisted upon and continuity should not be left to fortuitous contact.

15.7 I.E.E. wiring regulations

A typical example of how regulations affect the problem of earth fault protection can be seen in the interpretation of Regulation D.29 (I.E.E. Wiring Regulations) which requires that earth conductors shall have a resistance of no more than twice that of the largest current carrying conductor in the circuit. For many years it has not been economically possible for lead-sheathed, paper-insulated cables to comply with the regulation because the lead alloys now used have higher impedances than previously. The regulation is unnecessarily penal because protection is easily achieved by the use of fuses which are more sensitive than the regulations recognize. It can easily be shown that the use of H.R.C. fuses which comply with B.S. 88 would achieve the desired result with cable-sheathed conductance values as low as 12 per cent against the present requirement of 50 per cent. This is possible because Regulation B.23 stipulates the maximum volt drop and this in turn limits the length of cable or conductor and thus controls the limits of sheath impedance. Negotiations within the industry have been in progress for some years to remove these anomalies. The solution offered by H.R.C. fuses has long been recognized by all competent authorities and the need to change the regulations has been admitted. It is obviously not necessary to contemplate drastic changes in the construction of cable sheathing. In any event thousands of miles of cables which have been installed over the last twenty-five years are ostensibly non-compliant. The further alternative of providing more sophisticated protective devices is neither practicable or economic. In practice the situation is further mitigated by the fact that the cable sheathing forms only part of the earth path, particularly in an industrial system, where the practice of parallel bondings is common.

15.8 The general case

In the general case the choice of methods for earth fault protection lie between the use of a less sensitive but very reliable device which depends upon a reasonably well founded system and a more sensitive but intrinsically less reliable device which attempts to compensate for the shortcomings in the system. This is an over-simplification of the position, but is a reasonable approach towards an initial examination of the subject. It must be allowed that in some situations the simpler device is not sensitive enough and to recognize that in others the more complicated devices are too sensitive. The relative skills and knowledge of operating personnel is an important factor as is legislation relating to customer/supplier relationships. In the latter regard domestic installations are perhaps more critical than industrial.

H.R.C. fuses stand on their record as being the best overall proposition for the earth fault protection of most types of installation. When used in this context they are technically adequate, economically attractive and actuarially impeccable.

16 Fuses and fusegear enter new era

The issues facing fuse designers in the 1970s are as numerous as in any previous era. Economic pressures, world demand for materials and increased cost of labour all combine to reduce safety margins and to make the business of protecting electrical systems even more critical.

Industrial installations continue to grow in size and complexity; semiconductor devices have become larger and more versatile; high-voltage systems are increasing in the industrial sphere; accommodation for equipment is at a premium because of the cost of building space and newer devices such as vacuum interrupters are bringing new challenges and opportunities.

The move towards harmonization of standards within the E.E.C. and internationally further afield is a worthwhile objective but is creating its own problems. Rationalization in fuse design should not in itself be too difficult but it is not feasible to change fuses without regard to the equipment in which they fit and with which they are associated. Fuse rationalization cannot therefore proceed in isolation, and in the event must follow the harmonization of equipment design, while conforming to established codes of practice. There seems little doubt that fuse technology can keep abreast of any changes which become necessary once there is agreement on the primary issues.

Originally published in *Electrical Review*, July 1972.

Fuses and fusegear these days cover a very wide range of products and progress in one sector or another is always to be expected. New designs, in fact, continue to appear at a steady and vigorous rate.

Cost effectiveness in its widest sense is still the main objective of the new designs. This does not imply any diminution of quality; on the contrary the demand for improved standards of performance is as much in evidence as at any time in the past. This arises partly because of the need to maintain and, where possible, to improve safety standards in the changing environment and partly due to the fact that the protection of electrical services is becoming more vital as industry becomes more capital intensive.

There is considerable activity in international circles, particularly through the I.E.C. (International Electrotechnical Commission) and under the impetus of the E.E.C. towards the harmonization of codes of practice and standards. One manifestation of this activity has been the proliferation of international conferences relating to electrical safety and practice, all of which influence market demand and lead to changes in specification.

Load characteristics and load factors continue to increase. This intensifies the problems affecting heat dissipation in equipment and space saving as well as the general considerations relating to protection. Other and less direct factors are also significant, as for instance, the attitudes which have inspired the Trades Description Acts to improve consumer safeguards. These find relevance in the industrial sphere and thus affect electrical products, particularly those like fusegear which are in volume production.

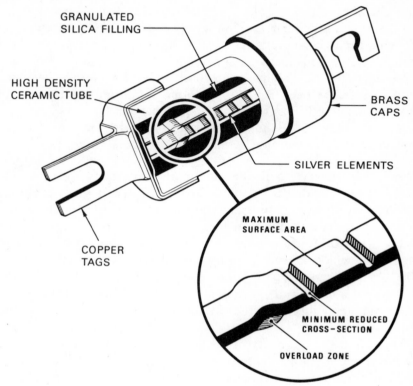

FIG. 16.1
Radical developments have been made in fuse element design. Some of the design features of 'GEC-English Electric' H.R.C. fuse-links are shown

In an electrical installation fusegear is one component along with switch-gear, transformers, cables of various types, motor control gear and a variety of consuming devices. Any change in any one of these components affects the rest. The changes in cable ratings and cable practice for example have directly affected fusegear; the new standards relating to motor control gear have required new fuse characteristics to achieve the desired co-ordination. While there has been little practical or technical difficulty in adapting fusegear to meet these requirements, it has been rather more difficult for the written regulations and standards to keep pace. The situation leads to some complication because the application of fusegear has to be reduced to fairly simple terms for the benefit of relatively uninformed users. The absence of up-to-date regulations can also restrict new developments.

There is still need for considerable discretion to be exercised by competent people in the choice of equipment for particular system requirements.

The main developments in standard industrial H.R.C. fuses have been improvements and variations in performance within the same package dimensions. This secures all the advantages of standardization which leads to economy of manufacture, but because these devices remain virtually unchanged in appearance the very considerable advances in design are not always realized. Rupturing or breaking capacities, time/current characteristics, relative accuracies, energy let-through and watts loss, etc., have all been greatly improved and the economic advantages derived have been considerable. The higher utilization of basic materials involving higher running temperatures has been achieved without compromising fuse performance, but perhaps the most significant improvements have been those affecting 'fault energy limitation'. This is becoming increasingly more important in the more complex industrial systems because it materially reduces 'down time' when faults occur.

The last year or so has seen the culmination of many years of efforts by Electricity Boards to standardize systems and to achieve logical co-ordination of protective devices throughout. Fuses are an important component in these systems and have had to be made to conform to a number of specification changes which have taxed the skills of fuse designers. One instance from many which illustrates these trends is that which concerns consumer service intakes. These are increasing in size and 100 A services are becoming the rule rather than the exception because of the increased use of storage and space heating. Increased duty on the fuse without any reduction in reliability of consumers' supply has called for considerable ingenuity.

Nothing has proved the versatility of the H.R.C. fuse more than the protection of solid-state devices and equipments. Solid-state technology is still in a state of rapid development and the market remains very fluid. As each new generation of diode or thyristor emerges new fuses have to be produced to match them. It is still true that the effective utilization of solid-state devices depends very largely on the degree to which they can be protected.

For those semiconductor devices which are already established it has been possible in recent times to consider the standardization of fuses. I.E.C. and B.S.I. are currently engaged on the task. Specifications governing these fuses are obviously more sophisticated than those for standard industrial fuses and new terms and technology have had to be adopted.

Beyond the fuses required for the established market are those needed to satisfy pioneer applications as new devices are developed and new uses found for solid state arrangements. The activity in this field is intense and involves more critical values of fuse performance in all parameters. The trends towards higher current and voltage ratings introduce new problems affecting speed of operation and let-through values but fuses have kept pace with requirements.

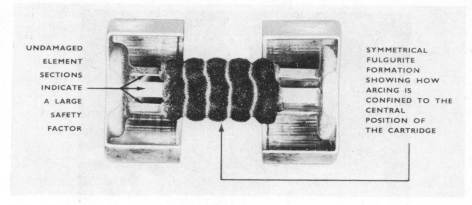

FIG. 16.2

Breaking or rupturing capacity of modern high-performance fuses is no problem. Shown here is the interior of a 'GEC-English Electric' 'blown' fuse after clearing a 200 kA fault

It is not usually possible for fuses to be specified until the design of solid-state equipment is in its final stages. Where new designs of fuse are found to be necessary there are then problems of resolving the commercial viability of any particular proposition. For this reason a good deal of development work has been put into producing fuses which can be made from standard components, but which are versatile enough to cover an enormous range of ratings and characteristics. These developments are well advanced and are already proving extremely fruitful. Further important progress is in prospect.

High-voltage fuses fall roughly under three headings serving the following applications:

(a) distribution systems;

(b) motor circuits; and

(c) other applications, such as the protection of instrument circuits, semi-conductor equipment, capacitors and other more specialized applications.

Electricity Boards at home and supply authorities abroad are using fuses to an increasing extent on systems ranging between 10 and 20 kV. Their demands have led to fuse development along the lines of obtaining better thermal performance, more precise characteristics, higher overvoltage capability and generally improved reliability. At the same time, a considerable degree of standardization has been achieved as a result of Electricity Boards crystallizing their views on system design and co-ordination and through the I.E.C. Distinction is drawn between 'general purpose' fuses which are built to perform safely over the complete band of overcurrents from the lowest overload condition to the highest fault condition and 'back-up' fuses which operate over a limited band of fault currents and are intended for co-ordination with switching devices having inherent overload capacity.

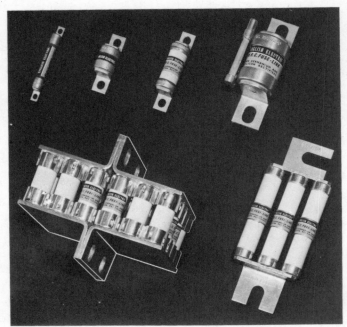

FIG. 16.3
*An extended range of fuses for protecting
semiconductor devices has been
developed by GEC-English Electric
Fusegear Ltd*

High-voltage fuses are available at all the preferred voltages and over a wide band of current ratings for use in air or in oil. Most have striker pin devices to actuate trip bars and achieve three-phase tripping when a single fuse blows. A good deal of ingenious development has been put into striker devices which may be specified to produce various levels of energy output. High-voltage fuses are used in the U.K. both in air and in oil, but Area Board demand for fuses in oil has tended to predominate in the last year or two. Consequently, a good deal of development has taken place, not only to make fuses reliable in oil but also to take advantage of the increased thermal ratings which are possible in this medium. Fuse package dimensions have been standardized to a considerable degree and most switches now available are able to accommodate the standardized fuses with obvious benefits in availability and interchangeability.

Co-ordination of characteristics between the fuse and other devices in the system, both up stream and down stream, has taken a number of years to resolve, mainly because of the need to reconcile new standards with the needs of existing systems. The developments in improving these characteristics to meet the new requirements have significantly increased the attractiveness of fuse protection and opened up many possibilities which hitherto might have been considered doubtful.

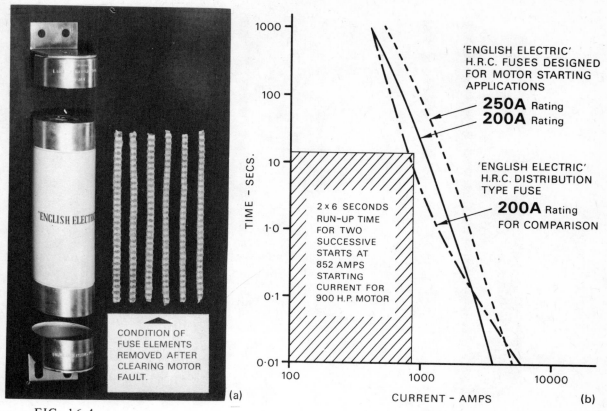

FIG. 16.4

(a) and (b) H.R.C. fuses for motor circuits can be made to withstand higher starting current than before. Smaller fuses can be used for given hp ratings; larger fuses can be used without compromising short-circuit performance

All fuses are voltage-sensitive and must be capable of interrupting safely at the highest voltages likely to be encountered in service. This has led to greater stringencies in the specification of voltage rating and created a demand for test evidence at voltages considerably higher than nominal system values.

The emergence of heavy-duty, high-voltage contactors, vacuum switches and M.S.D.'s (motor switching devices) in combination with fuses to take advantage of the considerable economic savings obtainable has stimulated a number of important new developments in fuse design. To be satisfactory in a motor circuit, the fuse must not only have adequate breaking capacity, but must also provide fault energy limitation to reduce the damage caused by faults and hence to reduce 'down time'. It must also be capable of precise co-ordination with other protective devices in the circuit. Motor circuit duty is much more critical as regards co-ordination than that needed for distribution systems because it is necessary for the fuse to conform to a

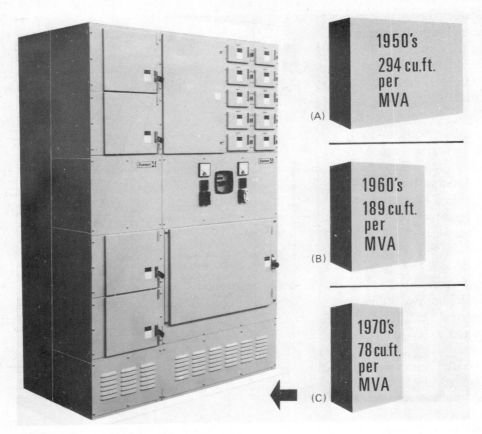

FIG. 16.5
*Substantial reductions in the size of fuse switchboards have been achieved since the
1950s. A modern example is the 'GEC-English Electric' System 4 fuse switchboard*

precise takeover point under fault conditions, while, at the same time, being able to
cope with the very onerous pulsing duty associated with motor starting.

'Pulse withstand' is not a new factor in fuse technology, but has become of vital
importance in relation to high-voltage motor circuit duty. It is essential to ensure
that fuses will not operate, except when called upon, to clear a fault beyond the
breaking capacity of the associated contactor or switch. It is for this reason that
fuses for this purpose need to be generously rated. Even this often leaves a zone of
uncertainty because, unfortunately, it cannot be guaranteed that the motor-starting
duty will remain consistent. Variations occur inadvertently and sometimes advert-
ently due to difficulties in maintaining operational discipline and also due to system
variations. In other cases the duty is necessarily onerous to meet the demands of
production.

For all these reasons it has been necessary to carry out intensive research and
development into improving the 'pulse withstand' of fuses and this has resulted in

successful designs now available on the market. These improvements have been achieved without compromising the effectiveness of the fuse in any other parameter. Indeed, increased rupturing capacities, faster action with reduced 'let-through' and smaller dimensions have been possible.

There has been no lack of incentive for the further improvement of fusegear equipment.

Vastly increased costs of buildings have focused attention on the cost of space required to accommodate electrical services and on switchboards in particular. The point is further accentuated by the fact that with increased use of electricity fuse switchboards are electrically much larger than hitherto. The number of circuits has increased and the operational facilities are more sophisticated.

Fusegear along with all other equipment to which cables are connected has been considerably affected by the drastic changes which have taken place in cable design and practice. The revision of cable ratings resulting from metrication and the introduction of new materials such as aluminium for conductors and plastic materials for insulation has tended to upset the thermal balance of the system and impose a greater burden on associated equipment. It has been necessary to accept either higher working temperatures in the equipment, or, alternatively, to redesign to improve heat dissipation except for those cases of better quality equipment which is already rated generously enough to avoid practical difficulty. The introduction of crimped connections has called for new concepts in terminal design, but these have not completely superseded plumbed and sweated joints. Consequently, fusegear equipment has to be made to accommodate both. For all these reasons, and also due to the need for smaller cabling chambers, it has been necessary to modify designs and amend specifications. This has coincided also with the demand for front access to switchboards to economise on labour costs on installation and maintenance.

Safety features such as the shrouding of live metal to safeguard unskilled personnel and the demand for 'on-load' switching for operational safety are in demand. Standards and regulations are reflecting these requirements to an increasing degree. Meanwhile, new materials are being introduced with a marked increase in the use of mouldings and pressings to achieve economy where volume production can be justified.

The trends towards rationalized and modular construction to reduce delivery times and improve general availability have not always been compatible with the demand for the degree of versatility normally associated with custom-built equipment. Attempts to combine the advantages of equipment which can be quickly assembled while retaining all the virtues of custom-built equipment have been a live issue with designers for some time. Other factors such as space-saving, which is now an economic necessity; operational performance and safety, which has increased in degree; service functions, which have become more complicated; and reliability, which remains of overriding importance, have also had to be fully taken into account. British equipments are already available which satisfy all these objectives to a substantial degree and new designs are forthcoming.

Equipments are now emerging which comply with new International Standards and in many cases anticipate trends on which future standards may be set. Harmonization in design and practice with those of all other countries and, in particular, with the E.E.C. still requires a good deal of thought and perhaps changes of attitude. The best British equipment is unsurpassed as regards performance and there is no reason to suppose that British concepts of construction will be found wanting in any respect when attitudes concerning standards of safety and criteria of utility and economy are finally resolved.

Index